SECULAR BIOETHICS IN THEOLOGICAL PERSPECTIVE

Theology and Medicine

VOLUME 8

The titles published in this series are listed at the end of this volume.

SECULAR BIOETHICS IN THEOLOGICAL PERSPECTIVE

Edited by

EARL E. SHELP

The Foundation for Interfaith Research & Ministry, Houston, Texas, U.S.A.

KLUWER ACADEMIC PUBLISHERS

DORDRECHT / BOSTON / LONDON

Library of Congress Cataloging-in-Publication Data

Secular bioethics in theological perspective / edited by Earl E.
  Shelp.
       p.   cm. -- (Theology and medicine ; v. 8)
    Includes index.
    ISBN 0-7923-3735-2 (hb : alk. paper)
    1. Medical ethics. 2. Medicine--Religious aspects. I. Shelp,
  Earl E., 1947-   . II. Series.
  R725.55.S43 1996
  241'.642--dc20   95-20651

ISBN 0-7923-3735-2

Published by Kluwer Academic Publishers,
P.O. Box 17, 3300 AA Dordrecht, The Netherlands.

Kluwer Academic Publishers incorporates
the publishing programmes of
D. Reidel, Martinus Nijhoff, Dr. W. Junk and MTP Press.

Sold and distributed in the U.S.A. and Canada
by Kluwer Academic Publishers,
101 Philip Drive, Norwell, MA 02061, U.S.A.

In all other countries, sold and distributed
by Kluwer Academic Publishers Group,
P.O. Box 322, 3300 AH Dordrecht, The Netherlands.

Printed on acid-free paper

Printed in the Netherlands

Table of Contents

EARL E. SHELP

Introduction

Theologians and theologically educated participants in discussions of bioethics have been placed on the defensive during recent years. The dominance of religious perspectives and theological voices that marked the emergence and establishment of "bioethics" in the late 1960s and 1970s has eroded steadily as philosophers, lawyers, and others have relativized their role and influence, at best, or dismissed it entirely, at worst.

The secularization of bioethics, which has occurred for a variety of reasons, has prompted some prominent writers to reflect on what has been lost. Daniel Callahan, for example writes, ". . . whatever the ultimate truth status of religious perspectives, they have provided a way of looking at the world and understanding one's own life that has a fecundity and uniqueness not matched by philosophy, law, or political theory. Those of us who have lost our religious faith may be glad that we have discovered what we take to be the reality of things, but we can still recognize that we have also lost something of great value as well: the faith, vision, insights, and experience of whole peoples and traditions who, no less than we unbelievers, struggled to make sense of things. That those goods are part of a garment we no longer want to wear does not make their loss anything other than still a loss; and it is not a neglible one" ([2], p. 2).

As a means to redress this loss and distinguish and justify their place in the forum of bioethics, theologians and theological ethicists have been challenged to identify and state more explicitly and persuasively the distinctive contributions theology can make to discussions of moral issues in medicine. An opening response to this challenge appeared in 1985 as a collection of essays titled *Theology and Bioethics: Exploring the Foundations and Frontiers* [6]. Process theologian, John B. Cobb, Jr., wrote in the epilogue, perhaps surprisingly, that theologians make "no contribution that in principle cannot be made by others. But that does not mean that they make no contribution at all, or even that they make no distinctive contribution" ([1], p. 306). As James M. Gustafson observed ten years earlier, much of the contribution of theology will depend upon what theologians claim about God and human moral agents related to God. More specifically, the contribution "depends upon

vii

E. E. Shelp (ed.), Secular Bioethics in Theological Perspective, vii–xiv.
© 1996 *Kluwer Academic Publishers. Printed in the Netherlands.*

whether the symbols or concepts of God provide a basis for drawing moral inferences with reference to human activity" ([4], p. 15).

Theological doctrines, moral teachings, world-views, styles of life, and other factors reflect certain understandings of God and God's will for human conduct. The diversity of understandings and the intensity with which they have been embraced account for the presence of several major religious traditions and divisions or groups within these traditions. Secular scholars may argue that this failure to have a common mind minimizes the perspectives or conclusions of theologians regarding matters in the secular realm of medicine. However, disarray among theologians and theological ethicists is not unique among intellectual disciplines. Alasdair MacIntyre, among others, has called upon philosophers to consider the diversity and disorder within philosophy [5]. Philosophers, despite their criticisms of theology and moral theologians, seem no more able to provide a way to decide between competing conceptual, methodological, and normative claims.

John Cobb's response to the complaint of secularists regarding the value of religious perspectives is instructive. He wrote, with respect to Christianity, "This does not imply that it is useless to ask theologians to address the issues. It does imply that one should not expect them to lay out Christian distinctives and derive from that in any direct way answers to the issues. To do that would misrepresent the nature of Christian faith. Also when Christians engage in the discussion of public issues, they should not be expected to refer often, if at all, to the story by which they live. This story has influenced their sensibilities, opinions, values, and commitments, but it is these, and not the story itself, that relate directly to the issue at hand. And these are not uniquely correlated with the story and its center. The efforts to establish too tight a connection distorts the faith" ([1], p. 306). Most of the essays in this volume illustrate Cobb's point. These contributors implicitly and explicitly disclose, in part, what difference there is between religious and secular thinkers. Moreover, each provides a critique of secular morality that, among other things, suggests what would be lost if religious voices were silent or absent.

As a whole, the essays in this collection represent a non-systematic critique of secular bioethics while illustrating, at the same time, conceptual, methodological, and normative differences and emphases within the broad disciplines of theology and theological ethics. The latter are more extensively reviewed in Paul Camenisch's edited volume on *Religious Methods and Resources in Bioethics* [3]. The essays in this volume are divided into two sections. Six essays in the first section broadly consider "Secular Inadequacies and Theological Contributions". The second section containing seven essays shifts the analysis to "Practices, Concepts, Methods, and Theories" in theological review.

The collection opens with Courtney Campbell identifying weaknesses in a solely secular bioethics. He locates religious interest in bioethics within the "pastoral" responsibilities of faith traditions and religion's concern with human meaning and purpose which bear upon practices of contemporary medicine. By relegating religious perspectives on issues in bioethics to the margins of public discourse, Campbell proposes that the substance and method of bioethics has suffered. More exactly, he argues that the purpose and meaning of medicine and health care have escaped rigorous examination. Without consensus about these substantive matters, secular bioethics forwards an understanding of medicine as a contractual relationship among self-interested parties who cooperate to advance their respective interests. He counters that a covenantal model of medicine, proposed by some religious ethicists, allows for the healing relationship to be located in a substantive context where purpose and meaning can be derived and from which an ethic and spirit of responsibility can be developed. Campbell argues that religious bioethics and medicine both are rooted in notions of the good for humans and their communities. However, the substantive influence of these notions in medicine has been lost, under the influence of secular bioethics, to a primacy of self-determination and individual choice. Sought after consensus has not been realized because substantive perspectives of the good continue to obstruct the path of proposed procedural solutions to dilemmas. Campbell argues a theological corrective to the limits of secular bioethics is marked by an ethic of responsibility and a spirit of humility. He concludes that discussions of public policy should be open to religious voices which supplement secular bioethics and provide ingredients obviously lacking within solely secular bioethics.

Gerald Winslow similarly thinks that part of the noble character of medicine would be lost if medicine is understood in solely secular terms. Winslow is concerned about how medicine is described and discussed. He supports his thesis by examining the interaction of metaphorical language and biomedical ethics. He reviews four metaphorical systems of talking about health care which describe and preserve distinct meanings of health care: religion (ministry of healing), military (war against disease), legal (context of patient rights), and business/marketplace (health care industry). Each metaphoric system interacts with certain moral values, judgements, and actions. Winslow warns against the withdrawal of religious language from the contest of competing health care metaphors. By preserving and protecting the religious metaphor of health care as ministry certain important values many be saved. He concludes, "The fundamental tasks of health care still require sincere, courageous compassion in the face of illness and eventual death. The language of spiritual ministry has supported such care in powerful ways. The preservation of such language is not impossible. And it is worth the trouble".

B. Andrew Lustig also argues against the idea that religious based arguments are irrelevant or unnecessary to policy discussions of health care. He considers the issues of health care reform and health care rationing in light of Roman Catholic social teaching. Lustig finds support within Catholic social teaching to support access to health care as a positive right. Moreover, he applies Catholic themes, principles, and arguments to proposals for health care reform and rationing to further demonstrate their relevance and distinctive perspective on the issues. He argues, in agreement with James Gustafson, that moral discourse may function in a number of ways and consist of several modes – ethical, prophetic, narrative, and policy. Debates about health care reform and rationing have tended to be ethical discourse, according to Lustig, which is too limited or restrictive. He proposes that other modes of discourse be included in order to expand the discussion beyond principles, including perspectives of moral significance formed and given content by religious values.

David Thomasma similarly finds enduring relevance and vitality within Roman Catholic moral reasoning for matters of moral concern occasioned by contemporary medicine and health care. He acknowledges that certain post-modern realities complicate understandings of moral authority, perceptions of moral truths, and reaching consensus about moral backgrounds. Nevertheless, "tradition" within Roman Catholicism still can inform moral authority with respect to issues in bioethics. He proposes that Roman Catholic moral analysis be forward-evoking, denying the temptation to fit present experience into past experience and past moral teaching. Thomasma argues the adoption of a new moral viewpoint that considers future implications and the context within which a policy is formed. He applies such a future experience viewpoint to Roman Catholic questions about reproductive technologies, euthanasia, and applications of the principle of gradualism to illustrate how bioethics and hermeneutics can be closely allied. Thomasma concludes that a realistic theological bioethics in a past-modern age entails a prospective hermeneutic tested in and by a believing community.

Somewhat like Thomasma, Robert Veatch's essay addresses points of contact of theological ethics with methods of moral analysis. Veatch considers the distress of some theological ethicists with Rawls's social contract which seems to result in humans inventing social norms. Veatch explains that Rawls's social contract process, when properly understood, leads to a discovery of norms. In short, "Rawls's social contract is essentially an epistemological metaphor for discovering the moral reality". This process is compatible with certain theological ethics that discover ethical norms. Moreover, Veatch suggests that Rawls may have disguised certain faith statements as premises that "end up shaping his principles and making them reasonably compatible with Judeo-Christianity". As such, Veatch implies that Rawls's techniques are not necessarily at odds with certain theological ethics with respect to issues in bioethics.

The discussion of moral principles becomes more focused in the essay by Karen Lebacqz. More specifically, Lebacqz examines the autonomy and beneficence paradigms of medicine. From a feminist theological perspective, she explores how compassion is an important base for medical ethics. Compassion, for Lebacqz, has two aspects – suffering with another and being impelled to counteract the suffering. Compassion involves entering into the experience of another and personally responding. This notion is developed by an explication of the Hebrew term for compassion which means "movement of the womb". Lebacqz considers this imagery important, in part, because of three functions or movements of the womb – receives, nourishes, expels. Drawing upon analyzes of Dorothee Soelle and Warren Thomas Reich, Lebacqz focuses on the aspect of reception as an ability to receive the pain of another into oneself such that it is experienced, not projected. It is this experience of connection to another, Lebacqz proposes, that guards against paternalism and the imposition of one's values on another. Thus, she concludes, compassion is needed in medicine to affect the perceptions and experiences, to avoid the trap of paternalism, and to enrich an understanding of beneficence.

The second section of essays on "Practices, Concepts, Methods and Theories" begins with a critique of clinical bioethics by Timothy Madison, a hospital-based theological ethicist and chaplain. Madison is concerned that bioethics in clinical settings is being reduced to a form or method of problem-solving suited for a secular environment of technical specialists and institutional policies. He thinks that James Gustafson's theologically informed concept of a community of moral discourse can, at least, complement secular approaches to problem resolution, if not move the discussion to deeper levels of moral analysis and character formation. Madison discerns in Gustafson's four dimensions of the work of a community of moral discourse in health care settings – data collection, open dialogue, continual reflection, and participation. Madison proposes that the concept of a community of moral discourse be introduced to clinical settings and the bioethical inquiry therein in order to place bioethics in a broader and longer term moral inquiry.

Stephen Lammers focuses on a concept commonly employed in clinical settings and referenced by academic writers in secular bioethics. Lammers contends that the content and breadth of the discussion of "futility" reveals a basic flaw in secular bioethics. What actually is at issue in these discussions, according to Lammers, is the "point and purpose of medicine". He traces the meanings and moral force of futility within secular bioethics to the prominence of the autonomy of persons and, secondarily, to justice or the allocation of resources. Noting that there is no consensus within secular bioethics about futility, Lammers suggests that theology can make a contribution to the discussion by raising fundamental issues of the point and purpose of medicine, including an explicit recognition of death, conceptions of a good life, the limited nature of medicine as a human enterprise, and the concept of care. He concludes that theologians and religious institutions of health care have roles to play in advancing these discussions which engage the basic issues of the point and purpose of medicine.

Another concept often employed in discussions of death and dying is evaluated from a Christian perspective by Michael Mendiola. Mendiola argues that the anthropology behind uses of human dignity is crucial to determining its content and force in aid-in-dying discussions. Much of the secular definition of human dignity is derived from the notion of "choice and control over the disposition of oneself and one's existential situation". Accordingly, honoring human dignity translates into accepting a person's choice and control. This understanding creates two problems for Mendiola with respect to the use of human dignity in aid-in-dying discussions: (1) Is death without choice and control robbed of dignity? (2) The emphasis on choice/control appears to reflect an individualistic anthropology and a priority for autonomy based upon it. As an alternative to the anthropology upon which secular understandings of human dignity are derived, Mendiola proposes a Christian theological anthropology that understands the human person and, thus, human dignity, in (1) relationships with God and other humans, and (2) wholeness, i.e., all the dimensions that constitute humanity. Honoring human dignity, from this perspective, involves more than accepting self-determining choices. It involves recognizing the many dimensions of human being and positive efforts to enhance and nurture them. Choice is necessary, but not sufficient, in honoring human dignity. The content of choices matters, as does the freedom to choose. According to Mendiola's theologically informed perspective, human dignity is not an a priori property of persons. It is an achievement of multi-dimensional human persons.

The theological critique of self-determination in death and dying debates is continued by Gerald McKenny. McKenny argues that the treatment of physician-assisted death in secular bioethics is too restrictive, depending too much on the notion of self-determination. He concedes that the traditional religious arguments against suicide made by Aquinas, Kant, and Augustine are not persuasive today. Moreover, distinctions regarding acts commonly found in discussions of this subject and certain consequentialist questions seem not to be serious moral hindrances to the growing consensus within secular bioethics in support of physician-assisted death. McKenny considers secular bioethics subject to criticism, however, with respect to its discussion of what makes "some self-determining choices better or worse than others". He addresses this issue by asking what responsibilities attend the medical and societal conditions conducive to a loss of meaning and worth near life's end. Drawing upon James Gustafson and Stanley Hauerwas, McKenny proposes that a broader understanding of moral duties linked to notions of the good is lacking in secular bioethics and leads to self-determination being enshrined. By premise and methodology, secular bioethics is driven to rely on self-determination and neglects the "practices that have made dying such a lonely and meaningless experience for so many persons". In contrast, McKenny concludes that theological approaches are more morally rich, complete, and satisfying.

The focus shifts from death and dying to abortion in the essay by Paul Simmons. Rather than secular bioethics being under scrutiny, Simmons critically examines the narrative ethics of Stanley Hauerwas to consider the potential of narrative ethics to advance moral discourse beyond principles, rules, and contexts. Simmons' critique centers on Hauerwas' personal construal of the Christian story, particularly certain assumptions considered problematic. Simmons identifies five pivotal assumptions for Hauerwas' approach to abortion: "(1) the narrative structure of Christian existence in the world; (2) the church as a community of character; (3) attitudes toward sex and childbearing; (4) sanctification in the Christian life; and (5) tragedy and suffering as signs of God's kingdom in the world". Several characteristics of Hauerwas' arguments and conclusions are considered highly sectarian and suspect by Simmons and other Christian ethicists. This dissent among Christian ethicists points to a need for more discussion of narrative ethics before concluding that Hauerwas' version and judgments are definitive.

Thomas Shannon's essay on "Genetics and Freedom" offers a critique of various claims of sociobiology to provide, in contrast, a philosophical/theological understanding of the reality of matter and human experience. Through an examination of the work of E.O. Wilson and Richard Dawkins, Shannon argues against materialistic interpretations of human experience. Drawing upon the thought on John Duns Scotus and others, Shannon challenges sociobiology's claim for a biological basis of all social behavior, the ascription of altruism as genetic selfishness, and determinism. Shannon concludes, "the images, metaphors, and arguments sociobiology uses are both flawed and inadequate in their attempts to capture the richness of human experience".

The collection concludes with an essay by physician-ethicist Robert Lyman Potter. His focus returns to a familiar concern throughout the volume – respect for the autonomy of persons. By using a comparative appraisal method of Irme Lakatos, he argues that a theocentric ground for respect for autonomy is more "acceptable" and "progressive" than a secular-humanistic one. Potter compares the thought of James Gustafson and H. Tristram Engelhardt to develop his thesis. While Gustafson's and Engelhardt's ideas find some general agreement, according to Potter, Gustafson's emphasis on "human fault" and its implications or effects is a point where the two part ways. Potter thinks that Engelhardt's methodology does not accommodate human fault in a way that does not limit the moral project of negotiating a peaceful agreement. To the contrary, Potter thinks that Gustafson's theocentric perspective for all human action addresses human fault and, according to Lakatos' test of rival theories, renders a theocentric model more "acceptable" and "progressive" for ethics.

The essays in this volume suggest that among theological ethicists there is an uneasiness and wariness of the substantive and procedural content of secular bioethics. While there clearly is no desire to reject secular bioethics in total, it is equally clear that these contributors think that something is missing in secular bioethics, that it is inadequate, not comprehensive, and, at points, non-responsive to the nuances of life that can become so morally troublesome. The tone and content of nearly every essay suggest the truth of John Cobb's notion that the sensibilities, opinions, values, and commitments of ethicists from faith traditions may be what is most distinctive. Perhaps it is at this point that theology makes its unique contribution to bioethics and constitutes a strong critique of the sufficiency of secular bioethics.

The dialogue between secular and religious ethicists about issues in bioethics appears to be expanding and intensifying. Theologians and religious ethicists increasingly are distinguishable from philosophers with respect to issues in bioethics. Moral perspectives grounded in and shaped by faith traditions are moving away from the margins toward the center of bioethics discussions once again. It appears that both religious and secular bioethics will benefit by a cogent and constructive exchange as the resources of each are subjected to critical examination. The essays in this volume constitute another step in moving this process forward.

Foundation for Interfaith Research and Ministry
Houston, Texas, U.S.A.

BIBLIOGRAPHY

1. Cobb, Jr., J.B.: 1985, 'Epilogue: Does Theology Make a Contribution to Bioethics?', in E.E. Shelp (ed.), *Theology and Bioethics: Exploring the Foundations and Frontiers*, D. Reidel Publishing Co., Dordrecht, pp. 303–307.
2. Callahan, D.: 1990, 'Religion and the Secularization of Bioethics', in 'Special Supplement: Theology, Religious Traditions, and Bioethics', D. Callahan and C.S. Campbell (eds.), *Hastings Center Report*, July/August, 2–4.
3. Camenisch, P.F. (ed.): 1994, *Religious Methods and Resources in Bioethics*, Kluwer Academic Publishers, Dordrecht.
4. Gustafson, J.M.: 1975, *The Contributions of Theology to Medical Ethics*, Marquette University Press, Milwaukee.
5. MacIntyre, A.: 1981, *After Virtue: A Study in Moral Theory*, University of Notre Dame Press, Notre Dame.
6. Shelp, E.E. (ed.): 1985, *Theology and Bioethics: Exploring the Foundations and Frontiers*, D. Reidel Publishing Co., Dordrecht.

SECTION I

Secular Inadequacies and Theological Contributions

COURTNEY S. CAMPBELL

Bioethics and the Spirit of Secularism

Bioethics emerged in the mid-1960s and early 1970s as an interdisciplinary field of study for a variety of reasons. One central cause was the development of new medical technologies that could create, sustain, and prolong life; biomedicine during this era was often criticized for "playing God", but it nonetheless is the case that the new technologies of genetics, organ transplantation, and reproduction, among others, collectively confronted human beings with responsibilities that historically have been attributed to the deity in western religious traditions. The necessary research and subsequent application of these technologies forced society to address and distinguish between moral and technological imperatives.

A second important factor concerned the social and cultural milieux of the period, which witnessed a protest and rebellion against traditional forms of authoritarian control, including the dominance of whites over blacks and other minority groups, and males over females. The result in both instances was the development of movements of liberation from traditional authority that expressed itself principally through the discourse of "rights", including civil rights and womens' rights. Traditional medical practice and authority was no less influenced by these movements of social protest and liberation. The dominant models of professional paternalism and the passive "sick role" of the patient were called into moral and legal question by a movement to recognize patients' rights to self-determination and autonomous decision-making in medicine. The Hippocratic ethos of patient beneficence, which promotes confidentiality but not honesty, is morally limited on issues vital to informed patient choice, such as truthful disclosure about diagnoses and prognoses.

It also became clear that the new medicine and its technology required decisions to be made about issues that transcended the intimacy and immediacy of the physician-patient relationship. The allocation of scarce resources, for example, kidney dialysis machines, required a different kind of approach than the dyadic, micro-ethic of traditional medicine. Indeed, the Hippocratic tradition did not address questions of justice and how a physician was to resolve conflicts of interest when there were more patients in need than available supplies of time or resources. A moral medicine would have to wrestle quite

3

E. E. Shelp (ed.), Secular Bioethics in Theological Perspective, 3–18.
© 1996 *Kluwer Academic Publishers. Printed in the Netherlands.*

seriously with matters of social and distributive justice, and do so with non-medical traditions of ethics, such as religion, law, or philosophy.

This ferment within medicine was, as Daniel Callahan [3] has argued, significantly influenced by religious ethicists and persons with religious interests. There are a couple of reasons why this should be the case. In a manner for which there is no philosophical parallel, faith traditions and religious communities have direct responsibilities to their members to provide practical guidance for situations in clinical medical practice. Thus, theologians addressing age-old questions of abortion, contraception, euthanasia, or suicide, as well as more recent issues of reproductive technology, genetic screening, or organ procurement, must be responsive and answerable to the pastoral needs of a faith-informed audience. By contrast, there is no such similar constituency addressed by philosophical writers in bioethics. To put the point somewhat differently, the issues of bioethics are necessarily more "real" and concrete for theological writers because they must address the lived experience of members of a faith tradition; it would be vocationally irresponsible to engage in fanciful abstractions based on presuppositions about an ideal or hypothetical world constituted by a homogenized "public".

At the same time, bioethics unavoidably gives rise to the profound questions of human meaning and purpose that are central to the teachings of the world's religions. The questions of the nature of the self and its origins; the nature of human nature in communal life; the human experiences of pain, suffering, dependency, diminishment; and the unavoidable prospect of death and human destiny are deeply embedded in the practices of contemporary medicine. For example, both the practices of life-prolongation through medical technology, and its recent response, requests by patients for assistance by physicians in suicide, presuppose particular understandings of the moral and existential significance of pain, suffering, and death. We may not engage in substantive debates over these matters for, it is claimed, a liberal pluralistic society is held together only by the glue of a social peace that comes from being agnostic about matters of the good life (or the good death). Nonetheless, religious approaches to bioethics cannot but consider these embedded issues as both moral and metaphysical.

In an insightful essay, LeRoy Walters has argued that religious thinkers and themes, particularly concerned with the relation of religion and science and religion and ethics, played the principal, and in some circumstances, the dominant role in a renewal of interest in medical ethics (and what would later become bioethics) between 1965 and 1975 [21]. His narrative does not imply that contemporary bioethics is historically dependent on religious considerations in any strict sense; significant contributions during these formative years of bioethics were also made by journalists, physicians, lawyers, and politicians. Nonetheless, it is fair to claim that the legions of scholars that now contribute to bioethics from many different disciplines are building on the foundational work of a prior generation of scholars for whom the influence of religious questions and ideas was substantial.

It should be noted here that this legacy is itself profoundly shaped by several institutional and intellectual developments in the 1960s that collectively culminated in broadening the ethical horizons of the religious traditions themselves. A new *ecumenical* tone surfaced in religious scholarship, one that was particularly on display in medical ethics, as religious ethicists were participants in dialogue with the moral resources and paradigms of other traditions. This was the case, for example, for Roman Catholic scholars working under the new openness to the world encouraged by the Second Vatican Council, while influential Protestant scholars, such as Joseph Fletcher and Paul Ramsey, gave serious attention to the values and challenges of Roman Catholic moral theology. This ecumenical character of religious discussion is an important forerunner of the subsequent interdisciplinary aspirations of bioethics.

When the venue of bioethics shifts, however, to what Richard J. Neuhaus has designated the "naked public square" [17], the relevance of religious reflection for bioethics may be vigorously disputed and contested, if not rejected outright. It is commonplace to hear or read arguments to the effect that it is inappropriate to appeal to religious convictions in a public forum because those convictions assume a substantive account of human flourishing that is not generally shared and in any event must not be "imposed" on others. A striking example from the writings of philosopher Dan Brock is illustrative of a consensus that religious discourse on bioethics has marginal significance, and should thereby become marginalized. In a very illuminating article on the merits of legalizing active euthanasia, Brock refuses to consider any perspectives on this question informed by religious positions, claiming: "[I]n a pluralistic society like our own with a strong commitment to freedom of religion, public policy should not be grounded in religious beliefs which many in that society reject" ([1], p. 19). Such a claim reflects, of course, but one view of the place of religious discourse in a pluralistic society, but it is an influential view, and it entails that the moral price for freedom of religion is that in matters of public policy, including those pertaining to bioethics, religious perspectives will be on the outside "looking in". It is, in short, appropriate for bioethics issues to be considered by a religious community based on the fundamental values and ethics of their faith tradition, and appropriate for religious ethicists to engage in cross-tradition scholarship and dialogue. A peaceable and tolerant discussion of the same issues in the public square, however, requires the silence of religious ethical perspectives.

I will subsequently want to take issue with this ideological tyranny of pluralism. However, I want first to consider the implications that seem to follow for religious discourse from the deference to pluralism. Will, for example, it make any difference for the substance and method of contemporary bioethics whether religious perspectives are marginalized or silenced? I want to propose a couple of historical analogies that lead me to give an affirmative answer to this question.

THE LOSS OF ENDS: AN ANALOGY

There are perhaps some understandable reasons why identifiably theological voices may in bioethics discourse blend into the background or fall silent altogether. James Gustafson has attributed the receding influence of perspectives rooted in religious thought to the requirements of pluralism for consensus in the midst of moral controversy, to "the effort to be persuasive on such grounds as diverse persons can agree upon", particularly in settings where an explicit theological appeal may become grounds for dismissing the practical conclusions of an argument ([9], p. 93). The question to be raised is whether, in the search for consensus, the hiddenness of religious themes entails a loss of any substantive significance for bioethics.

The first analogy I wish to propose is suggested by the relation of theological thought to the ethics of political statecraft and war in western culture, particularly as developed in the moral tradition of "just war". The historical evolution of this tradition drew on several sources, including classical philosophy, canon law, and codes of chivalry, as well as theological thinking rooted in biblical teachings and narratives and tempered by the realities of political power and statecraft [12]. Yet, during the medieval period when these various sources coalesced into a fully-developed moral tradition, religion was not only a shaper of the morality of war, but also often a cause of war. Against the background of the age of religious wars and the emergence of new political arrangements among newly-founded European nation-states, explicit theological discourse on moral reflection about war substantially diminished between the 17th and 19th centuries. The ethics and laws of war became the province of philosophers and proponents of international law, such as the Dutch jurist Hugo Grotius, who sought to establish consensus on the justification and limitation of war based on natural and positive law.

Correlatively, however, the moral tenor and character of the just war tradition was transformed. The philosophical and legal focus concerned the conduct of war, its regulation and control, and the moral and immoral *means* by which war might be fought. Yet, this necessary concern tended to displace an equally important and indeed preliminary question about the *ends and purposes* of war and overlooked the moral need to inquire critically about the moral authorization of war in the first place. The secularization of the just war tradition did produce a moral paradigm for discourse about war that was very attentive to the details of the conduct of war, but lost critical distance, or stated more theologically, lost a sense of prophetic criticism that challenged the presuppositions of war as a moral practice for human beings.

I want to highlight two significant lessons from this story of secularization of a moral tradition for the nature of bioethical inquiry, and the place of religious reflection in that inquiry. Many scholars have called attention to the pervasive military metaphor of medicine. William F. May, for example, has described the "physician-as-fighter" as one of the basic professional and cultural understandings of medicine [16]. Other scholars have observed that the language of war is deeply embedded in medical discourse, with some very significant implications for medical priorities and practice: "The military metaphor tends to assign priority to health care (especially medical care) over other goods, and within health care, to critical interventions over prevention and chronic care, killer over disabling diseases, technological interventions over care, and heroic treatment of dying patients" ([5], pp. 30–31).

It is precisely with these kinds of issues that the limits of a secular bioethics are revealed, just as they are with a secular tradition of war and statecraft. What is required to provide a more balanced medical care is a conception of the ends and purposes of medicine to complement the technological and invasive means of contemporary medicine. What currently serves as the paradigm goal of medicine is a conception of death as the "enemy" that must be defeated through constant "wars" on various diseases and substantial reliance on the technological "armamentarium" of modern medicine. But, this fosters a conception of total medical warfare, and simply does not allow for the kinds of discriminating distinctions needed to establish priorities among kinds of care, kinds of diseases, limitations on the technological imperative, and an understanding of death that would restrain overtreatment of the terminally ill. In this respect, then, a secular bioethics that presupposes an image of medicine as warfare suffers from the same kind of moral myopia that a secular ethical tradition of war does in general: There is a pronounced mismatch between means and ends that culminates in a moral distortion of the practice, be it medicine or war, or medicine as a metaphorical form of warfare.

There are recently signs of renewed attention among some bioethicists to questions of ends, purposes, and a moral teleology of medicine [4], [14]. These signs suggest new metaphors or paradigms for contemporary medicine, and it is not surprising that this renewal has been deeply influenced by questions of metaphysical meaning and purpose. Yet, if we ask why this discussion is still carried on in rather limited circles, the moral tyranny of pluralism again asserts itself. For talk of substantive ends and purposes of medicine, which can then provide the philosophical basis for priorities and limitations within medicine, unavoidably requires some sense of the good human and the good professional life. Yet, the language of the "good", while quite central to religious traditions and faith communities, is inadmissable under the moral logic of pluralism.

One suggestion from the historical analogy related above, then, is that religious perspectives can offer different and instructive metaphors for the ends and purposes of medicine, such as "healing" or "caring" in contrast to "curing". A second lesson also seems embedded in the story of secularization of a moral tradition. My argument does not intend to overlook the important and necessary attention devoted to the details and methods of medical practice in secularized bioethics, such as informed consent or research protocols, or advanced directives and required request legislation, which have expanded patient choice and enhanced protection of patients from harm. In these respects, substantial contributions have been made to a more moral medicine. Moreover, these practices have become institutionalized within the health care establishment through hospital ethics committees, ethics consultation teams, institutional review boards, and other regularized forms of ethical oversight of medicine. This oversight certainly requires a common meeting ground, in which the norms and principles of secular bioethics are integral. When bioethics attends to the clinical level of medical practice, and practical conclusions or decisions must be made *and justified* to a public audience, then secular bioethics is essential and necessary. It is not, however, sufficient.

Albert Jonsen and Stephen Toulmin, who served in the mid-1970s as staff philosopher and staff consultant, for the first national regulatory oversight board for medical ethics, the National Commission for the Protection of Research Subjects in Human and Behavioral Research, have commented in some detail on how members of this interdisciplinary commission were able to reach consensus and offer detailed policy recommendations on very difficult issues, including the ethics of research with fetuses, children, prisoners, and the institutionally mentally impaired [13], [19]. Yet, Jonsen and Toulmin observed a pronounced disparity between consensus on practical conclusions and theoretical disagreement that subsequently led them to advocate a casuistical method of moral reasoning to supplant principlism in bioethics. The crucial notion is consensus on the practical ethical choice facing the patient and physician, or the policy maker, not the principled-based argument for that choice. For, on Toulmin's account, "such principles serve less as foundations [for particular ethical judgments] . . . than they do as corridors or curtain walls linking the moral perceptions of all reflective human beings with other, more general positions – theological, philosophical, ideological, or *Weltanschaulich*" ([20], p. 32).

Toulmin's claim has initiated a decade-long debate in bioethics over the adequacy and limits of a principlist method. My concern here, however, is with the implications of a neglected aspect of his position, the philosophical dismissal of world-views. If consensus on controversial, at times, potentially polarizing, issues in bioethics can be achieved without recourse to normative principles, it follows on Toulmin's account that bioethics has no need either for what the principles reflect, namely, metaphysical positions, including religious convictions and theological world-views. That is, the purpose of bioethics is portrayed as achieving agreement, consensus, and closure; any ideology that holds out the prospect of divisiveness – and we have long historical experience that bears out that theological positions rooted in particularized conceptions of the human good promise exactly that – is for that reason to be viewed as morally suspect and marginalized in moral discourse on medicine.

The problem, however, is that agreement and consensus then become ends in themselves; secular bioethics cannot offer a critical standard by which we can morally evaluate the kind of agreement. The lesson to be learned from the historical analogy can now be made more explicit: We approach medicine as though we were signing a peace treaty (that is, an agreement to eradicate divisiveness) without any criteria to evaluate whether the agreement will produce a "good" or "just" or even lasting peace. And if anything, it should be clear from the history of war that just any peace will not be adequate to the human realities at hand, the Versailles Treaty to "end the war to end all wars" being the most notorious case in point.

As Stanley Hauerwas has observed, insofar as bioethics devotes primary attention to the resolution of dilemmas in clinical medicine, such as terminating medical treatment or truthtelling that may cause harm, "we tend to underwrite the practice of medicine as we know it, rather than challenging the basic presuppositions of medical practice and care" ([11], p. 71). These basic presuppositions include such matters as the origins, nature, and destiny of human beings; how medical technology can contribute to or diminish the dignity of the human person; the moral significance of pain, suffering, and dependency and their implications for medical practice and the cultural ethos; or the quest for meaning and purpose in dying and death. These issues are embedded within the very practice of medicine as an art that seeks the healing of the whole person. Thus, the divorce Toulmin proposes between the practical and the metaphysical makes especially little sense in the context of medicine: what will remain as the content of secular bioethics is a very truncated morality that will address a very depersonalized medicine. This is the moral price of consensual pluralism. It is a price of moral hypocrisy that should remind us of the criticism of Jesus directed at his opponents: In bioethics, we have "neglected the weightier matters" (*Matthew 23:23*), including justice, dignity, embodiment, and meaning in our (necessary) attention to the procedural means for a more moral medicine.

BIOETHICS AND THE SPIRIT OF SECULARISM: A SECOND ANALOGY

The secularization of the moral tradition of just war is but a specific instance of the general secularization of western culture. My analogy with contemporary bioethics has suggested that in demarcating the public square of bioethics discourse as the domain of secular bioethics, we risk the loss of critical distance or prophetic insight about the moral point and purposes of medicine. I want to elaborate this idea further by invoking a second historical analogy about the shift from religious to secular moral perspectives.

One classic account of the general process of secularization in the European west is found in Max Weber's *The Protestant Ethic and the Spirit of Capitalism* [22]. Weber contends that the commercial and contractual ethos that is embodied in the modern world paradoxically has its historical roots in religious thought, particularly the emphasis within reformed Protestantism to transform the world into a theater of God's glory. Weber argues that the quest for meaning and salvation (which he elsewhere describes as a distinctive and universal feature of the human species) had an enormously influential psychological and moral sanction on a person's "calling" in the world, particularly as displayed in a desire to excel in one's occupational vocation. The legacy of this religious root is still present in contemporary society (Weber's of the early 1900s and our own of the late 1900s) through a "work ethic". The work ethic reflected a sense of mission to conform the world to the ideal of the divine kingdom, while having the incidental effect of allowing persons to accumulate vast capital and wealth.

The Protestant work ethic thus is given a point and purpose through its setting within a religious world-view. However, through four centuries of secularization, given impetus by the Industrial Revolution and the philosophical Enlightenment, the practical ethic of work is severed from its metaphysical rootage. Through this process, as exemplified by Benjamin Franklin, the work ethic loses its "spirit" or foundational purposes within a theological world-view, and the acquisition of wealth becomes a value of intrinsic worth. The problem is not the loss of the ethic, because it survives, but rather the loss of the theological context that gave meaning and spirit to the ethic. One consequence of the waning away of these religious roots in western culture is a world deprived of moral meaning – the deep questions about the origin, nature, and destiny of human beings and of their communities gradually become lost from view and may no longer be asked at all.

For Weber, the triumph of secularization is a cause of lamentation rather than celebration, to the point that in the concluding pages of *The Protestant Ethic*, he drops pretenses to value-free sociological and historical description, and adopts a critical, normative perspective on these developments. We now live in an "iron cage" because we have appropriated and live out an ethic that has lost its moral direction, and there are no compelling alternatives other than to make financial gain a good in and of itself. This in turn has a corrosive impact on moral character, for as a people, we have become "specialists without spirit; sensualists without heart" ([22], pp. 181–182). We have lost a personalism in our relations because personal self-interest has become the overriding "spirit" of the work ethic. We thereby find ourselves enclosed in a cage in which genuine human relationships are no longer possible, and a social ethic of moral strangers prevails.

Although the particularities of Weber's account continue to be much disputed, I want to propose that his comments about the process and outcomes of secularization are particularly insightful for understanding contemporary bioethics and its secular spirit. The analogy I wish to explicate here contains several features.

In a fundamental way, the nature of the health care professional's calling and his or her relationships with patients have undergone a profound paradigm shift under the auspices of secular bioethics. The inegalitarianism of the paternalistic ethos of the Hippocratic tradition has been morally discredited for legitimate reasons. The paternalistic model presumed a "doctor-patient" relationship in which, to use the etymological roots of the terms, the doctor was the one who taught, and the patient the one who suffered. There was, to be sure, within the Hippocratic tradition a moral purpose and context for paternalism, namely, to benefit the patient, but the patient's welfare tended to be defined by the professional.

The contemporary relational model in bioethics has drawn on sources in law and business to rectify the inequalities of paternalism. This is typically expressed through the metaphor of "contractual" relations, in which the contracting parties are portrayed in a relationship of "provider-consumer". The assumption in the contractual model is that two self-interested parties can cooperate to further mutual interests. The power in this relationship essentially resides with the "consumer" who can demand that his or her needs be provided for through the language of rights and entitlements. If we then inquire about the moral context or purpose of the professional's calling in this relationship, it becomes clear that it is the pursuit of individual interests, whether those of the provider or the consumer. This is the "spirit" of contemporary secular bioethics. That is, the contractual relationship is coherent only within a context of moral strangers who have no shared substantive values (although they may share procedural values of cooperation, negotiation, and accommodation).

It is, however, no surprise that litigation has flourished under this model since its conceptual structure and language presume depersonalization. This in turn gives us insight into the theoretical and practical inadequacies of the secular contract. A person cannot encounter a physician in a position of consumer, because if illness is present, the option of not purchasing services is foreclosed. In short, the ill person is vulnerable and dependent in a way not captured by the consumer metaphor. The physician experiences, moreover, a profound conflict of loyalties: A tendency to restrict one's professional activities to contractual terms leads to "moral minimalism" [2]; a fear of the law ensues in over-testing, over-treatment, and medical maximalism. Most significantly, professional deference to the demands of individual choice means that the moral goods constitutive of the calling or vocation of medical practice have diminished to the point of irrelevance. One sign of this is the profession's unwillingness to take a moral stand on controversial moral questions, such as physician-assisted suicide. In the midst of a 1994 citizen referendum on this issue, the president of the Oregon Medical Association recommended a stance of "neutrality" because physicians need to hear what their patients and the voters had to say on the question [18]. It is difficult to conceive of other professions that have so readily given up their authority and moral responsibility. Nonetheless, the spirit of individual choice that drives contemporary bioethics has transformed the physician's calling into the technical "specialist" that Weber envisioned.

It is not that this transformation and secularization of medicine is necessarily inexorable. Religious ethicists have long affirmed an alternative to both paternalism and contractualism that is constructed around the idea of "covenantal partnership" [15]. Equality in the relationship is affirmed by an understanding of a shared and reciprocal dependence between patient and physician. The covenantal model preserves the idea of medicine as a moral vocation with certain basic goods, including but not limited to respect for personal autonomy, as constitutive of medical practice. Partnership also affirms certain characteristics of gift relations, rather than the legalistic minimalism of contractualism, such as a mutual acknowledgement of "responsibilities" in the relation to correspond with affirmations of individual rights.

The theme of covenant reflects a substantive theological context within which the healing relationship is located and from which it receives purpose and meaning. This context draws on the formative covenantal narratives embedded in the western monotheistic traditions. Its content involves the divine "calling" of a people to responsibility, a responsibility that requires embodying the divine character in the world by a people who have historically undergone slavery, marginalization, and oppression. This is the revolutionary reversal proposed by religious ethical traditions, that the last in rights are called to be first in responsibility through the liberating covenantal relationship.

A theologically-informed view of the clinical relationship in medicine will offer both an ethic and a spirit of responsibility that transcends the individualism of contract on one hand and the tyranny of Hippocratic paternalism on the other. It provides a substantive view of the good of the healing relationship while assuming that this is a mutual and shared good rather than one unilaterally imposed or projected by the more powerful party. For the check on power is the call to responsibility of the dependent and least powerful. The covenantal vision of partnership cannot accommodate the passivity embedded in the "sick role" of a patient. Nor, however, will it accommodate a domineering ideology of individualism to corrode and diminish the moral vocation of medical practice to technical, dispassionate specialization. In the realms of moral character and metaphysical meaning, then, religious bioethics can both critique and renew the secularized specialist without spirit or the sensualist without heart.

A THEOLOGICAL CORRECTION: THE RESPONSIVE SELF

I have claimed that religious bioethics is rooted in a substantive understanding of the good for human beings and their communities. Secular bioethics, by contrast, is agnostic about the good so that, in a pluralistic culture, consensus may be forged in the public square among persons who meet as moral strangers, that is, as persons with different, and often conflicting perspectives of good and meaning. The historical tradition of medicine has also been more substantive than procedural, with certain inherent goods and values constitutive of medicine as a moral practice. Yet, under the influence of secular bioethics, with its more restricted moral horizons, the good of medicine has been transformed to the pursuit of one value, the enhancement of self-determination and individual choice. While in theory, self-determination should apply to both parties in a medical relationship, in practice it frequently has the effect of reducing the physician to a technician who provides what the medical consumer wants. I want to inquire about the significance of this loss of moral particularity in medicine for both religious and secular bioethics.

H. Tristram Engelhardt has argued that the "success" of bioethics in a secular, pluralist society requires that moral arguments be generally accessible and compelling to persons irrespective of their personal, religious, or professional moral commitments. In the dominant stream of philosophical bioethics, therefore, "generality is purchased at the price of content", and not surprisingly, given the historical analogies examined above, this deprivation of content is displayed precisely in an incapacity to address issues of meaning and purpose: "Philosophical ethics cannot supply a deep meaning to the suffering, debility, and death of patients" ([7], p. 88).

Secular bioethics gives us a world with minimal moral meaning; what meaning is present is dependent on personal invention or creation. We can be content with a bioethics deprived of meaning to the extent that the horizon of bioethics is limited to solutions to practical dilemmas. Problem-solving methods for the moral dilemmas of medicine require agreement and consensus, whereas history gives us reasons for being suspicious of substantive moral positions, especially those expressed in religious traditions, as divisive and polarizing. We may gain a substantive ethic by looking to such moral resources, but here the trade-off that concerns Engelhardt is reversed: Moral substance is gained at the expense of generality. In a pluralistic society, this is a most problematic gain, and one that could erode the success of secular bioethics to forge agreement among moral strangers.

If we focus, however, on areas where secular bioethics fails to be generally persuasive or achieve social consensus (for example, priority setting for the allocation of health care resources), we will find a bioethics deprived of substantive meaning considerably less appealing. The virtue of secular bioethics resides in its general appeal and capacity for moral closure, but if questions of a broad social scope cannot achieve consensual agreement, the loss of meaning and substance as the moral price for public acceptance begins to look like a vice. Indeed, the fact that we continue to experience vigorous moral disagreement in bioethics ought to reveal to us that secular bioethics is seriously suspect insofar as it holds out the promise of consensus via general accessibility. In Kuhnian terms, disagreement is the "anomaly" in the secular paradigm that requires the articulation of new ethical models or paradigms. Disagreement suggests that the presumed divorce of the moral and the metaphysical has not occurred, but rather that substantive perspectives of the good are obstructing the path to social consensus. However, examined closely, it should not be surprising that the individualist spirit of secular bioethics cannot deliver on its promise of moral closure for society as a whole.

A theological correction is needed to remedy the limits of a secular bioethics in a pluralistic society. This correction needs to begin with a different account of the self and human nature. The moral self is not the mathematical calculator required by utilitarian-inspired bioethics, nor the dispassionate legislator of universal law presumed in Kantian deontological ethics. Indeed, we might have difficulty recognizing dispassionate calculators as "human" agents in any meaningful sense. Rather, the moral agent is both finite and fallible: We sometimes are unable to determine the good, and sometimes incapable of doing the good we know. In more explicit theological terminology, secular bioethics needs to recognize and accommodate the tendency towards "contraction" ([10], pp. 305–306), of the inward turning of the self, endemic to all human beings.

A second element of this necessary correction is that secular bioethics needs to experience what I shall refer to as an "enlightenment of a Copernican revolution". That is secular bioethics presumes that moral choices revolve around the individual who creates his or her own moral universe. The agent of secular bioethics exhibits power, control, and domain over the moral realm. I am not persuaded that religious bioethics has fully articulated the shift in perspective and practice required by a biocentric or theocentric ethic, but at least the following points can be made. Religious bioethics must display a responsive ethic because the background conditions of our moral existence are given to us, including our genotype, and our natural, familial, and social environments. These create responsibilities and an ethic of intimates that we do not choose, for example, the moral ties of family and kinship, a context in which the moral language of "rights" is odd if not altogether inappropriate. In short, the world-view that informs religious bioethics is that we are responsive agents, subject to powers and forces beyond our control and power. We may experience these external powers as arbitrary and cruel, or creative and healing, but they no less shape the choices we have in medicine than do our internal capacities of rationality and emotion.

This paradigm shift should cultivate in the moral agent a disposition of humility, rather than the hubris expressed in secular culture. The experience of powers beyond our control requires a corresponding acknowledgement of human limits, even to the biomedical technologies that we often frequently look upon to provide deliverance from these powers, such as the inexorable biological process that moves us into a period of increasing frailty, diminishment, and death. However, humility requires that we understand the experience of dependency as central to our experience, not as some aberration in a culture that so highly values independence. We are, in fact, always "in-dependence", upon the powers and forces beyond control for life and sustenance; upon other persons for nurture and care in our formative years; and upon significant others for self-understanding and identity, friendship, and intimacy. Humility involves a realization that our moral world is made for us as much as we make it; the practical corollary of this is that we seek human flourishing through these relationships of dependency, rather than seeking to avoid them through technology, which is the world-view of the Weberian sensualist without heart. The virtue of humility therefore is the spirit of religious bioethics.

RELIGIOUS BIOETHICS AND THE TYRANNY OF PLURALISM

If an ethic of responsibility, rather than rights, and a spirit of humility, rather than individualism, characterize religious bioethics, we must now ask what status this ethic has in the public discourse of a pluralistic society. I do not propose here that religious bioethics supplant secular bioethics in public policy, but rather that religious bioethics can supplement secular perspectives without threatening the collapse of moral civility.

It first must be noted that pluralism offers a substantial benefit to religious bioethics. The principle of tolerance and respect for difference that issues from pluralism allows a cultural space for the articulation of theological perspectives and their practice within particular faith-traditions. Religious bioethics receives moral legitimacy and credibility from the foundational principles of pluralism, and therefore cannot be dismissed as irrelevant and parochial. Indeed, whatever pluralistic principle is applied to religious bioethics must also be applied to ethical traditions in the healing professions, including medicine and nursing, which no less embody certain substantive moral commitments.

A provocative illustration of the moral credibility of religious bioethics under the auspices of pluralism is presented by Ronald Dworkin, in his recent book, *Life's Dominion* [6]. Dworkin seeks to support a liberal conclusion about the right to abortion or the right to euthanasia by starting from conservative premises, namely, the religious appeal to the sanctity of human life. If we understand the debates over abortion or euthanasia, surely among the most contentious in our society, as fundamentally involving interpretations over one primary value, the sacrality of life, then Dworkin contends we will be driven to permit freedom of choice on abortion or euthanasia, not on the basis of secular values such as privacy rights or liberty interests, but rather from the theological tradition of respect for freedom of conscience. While I do not share Dworkin's substantive conclusions on abortion or euthanasia, his method insightfully illustrates how pluralism can integrate and legitimate religious perspectives and values in bioethics.

Pluralism is then in part parasitic on traditions of religious ethics, and this raises the paradoxical question of how a secular paradigm can survive absent its theological rootage. The historical analogies developed above have shown that such paradigms can survive, but only in minimalistic, truncated forms, because they lack a substantive vision of ends and meaning. Thus we can say in bioethics that individual choices and mutual agreements should be respected so long as they do not pose risks of harm to others. However, it is a very barren and sparse understanding of the moral life that limits our responsibilities to not harming others. We want as well some grounding for duties to benefit and assist others, especially as those are central to the moral identity of professional caregivers. However, a secular bioethics cannot provide such a source of moral obligation; we must instead rely on non-secular sources, such as religious traditions, or internal professional traditions, of ethics.

I do not wish to claim that pluralism's own reliance on religious, or at least substantive, sources, of ethics for its plausibility resolves issues about the influence and scope of theological perspectives in policy debates. On this matter, however, it is vital to distinguish between the *process* and the *outcome* of institutional or public policy in bioethics. I would affirm that the pluralistic parameters of our society are such that a public policy outcome or conclusion ought to be articulated and justified on such grounds as persons with diverse and conflicting moral visions can agree on. This means that secular bioethics, in both its procedural methods and substantive norms, will play the dominant role in justifying moral conclusions. It does not, however, sanction ignoring theologically-informed voices in the policy-making process. Rather, pluralism requires an open dialogue with such voices and respect and tolerance for views with which one may disagree.

At the very least, religious perspectives will assist in moral discernment, that is, our capacity to "see" moral problems, which is a necessary pre-condition of "solving" a dilemma. In many policy contexts, moral discernment can be enhanced through a narrative construction of the problem and analogical reasoning. For example, the biblical narrative of the good Samaritan has been appropriated in several influential secular bioethics discussions to differentiate minimally required moral action from moral heroism, or to demarcate legal from moral responsibilities. The narrative provides a basis to explicate the meaning, assumptions, and limits of the norm of beneficence. Its use suggests that what may be generally accessible to persons are not principles and rules derived from some moral theory but a shared heritage of stories and narratives. These stories do not provide a direct answer to the dilemma, but they do provide an interpretive context.

Moreover, a sustained and honest dialogue will illustrate how very difficult it is to divorce a moral position, religious or secular, from some kind of substantive cosmology or understanding of human nature. A secular position in bioethics that supports euthanasia presumes a conception of what it means to die with dignity no less than does a religious position that is opposed to euthanasia. If Toulmin is correct that moral principles reflect various world-views, it seems more candid and more in accord with the ethical requirements of pluralism to integrate those world-views into the policy process rather than partitioning them.

The necessity of world-views, of different ways of moral "seeing" and discernment, is perceived in almost all scholarly disciplines. A world-view, whether it be religious or scientific, political or economic, provides the basis for rebellion and protest against the status quo, and if sustained over time, the possibility of revolutionary transformation [8]. It was a different view of the moral world of medicine, and particularly the role of patient as a *person* in that world, that helped launch the incipient movement of bioethics. The problem now is that secular bioethics provides little moral basis for moving beyond individualism to examine broader claims of social justice. The critical distance and prophetic insight of religious world-views are especially important in challenging institutions and persons with vested self-interests in maintaining the practices of the current health care system, and through those expressions of self-interest, collectively dismantling any prospect for substantial reform of the current system, as occurred in the United States in 1994. Prophetic protest is needed to offer a sharp critique of entrenched institutional and financial interests, of a culture so idolatrously enamored of high technology medicine that it allocates health care resources inequitably, and of visible and subtle forms of discrimination against persons and groups that have historically experienced marginalization. Indeed, the historical practices of care and hospitality to the stranger by religious communities reveal that justice requires moving beyond a rationally self-interested egalitarianism to a commitment of special preference for the health care needs of the socially vulnerable and voiceless.

TRANSFORMING THE ETHIC AND SPIRIT OF BIOETHICS

I have tried to articulate the significance of the process of secularization in bioethics, through both historical analogy and contemporary example. Both lines of inquiry lead to a similar conclusion: Secular bioethics and its spirit of individualism is incomplete in theory and inadequate experientially without the goad and the vision of religious world-views. Those world views sustain an ethic of responsibility and a spirit of humble in-dependence that provide a necessary counterpart to the secular ethic of rights and self-determination. They offer purpose and moral integrity to a vocation – caregiving through medicine and nursing – that runs the risk of being reduced to a technocratic specialty and entrepreneurial enterprise. Perhaps most significantly, religious world-views promise an eschatology of hope in the face of cultural cynicism and despair. In order to transform health care delivery to conform to fundamental values of social justice and human dignity, bioethics must re-claim the legacy of liberation from which it arose.

Oregon State University
Corvallis, U.S.A.

BIBLIOGRAPHY

1. Brock, D.: 1992, 'Voluntary Active Euthanasia', *Hastings Center Report* 22, 10–22.
2. Callahan, D.: 1981, 'Minimalist Ethics', *Hastings Center Report* 11, 19–25.
3. Callahan, D.: 1990, 'Religion and the Secularization of Bioethics', *Hastings Center Report* (Special Supplement) 20, 2–4.
4. Callahan, D.: 1990, *What Kind of Life: The Limits of Medical Progress*, Simon & Schuster, New York.
5. Childress, J.: 1984, 'Ensuring Care, Respect, and Fairness for the Elderly', *Hastings Center Report* 14, 27–31.

6. Dworkin, R.: 1993, *Life's Dominion: An Argument about Abortion, Euthanasia, and Individual Freedom*, Alfred A. Knopf Publishers, New York.

7. Engelhardt, H.T., Jr.: 1985, 'Looking for God and Finding the Abyss: Bioethics and Natural Theology', in E. Shelp (ed.), *Theology and Bioethics: Exploring the Foundations and Frontiers*, D. Reidel Publishing Company, Dordrecht, The Netherlands.

8. Gunnemann, J.: 1979, *The Moral Meaning of Revolution*, Yale University Press, New Haven.

9. Gustafson, J.: 1975, *The Contributions of Theology to Medical Ethics*, Marquette University Press, Milwaukee, Wisconsin.

10. Gustafson, J.: 1981, *Ethics from a Theocentric Perspective*, vol. 1, University of Chicago Press, Chicago.

11. Hauerwas, S.: 1986, *Suffering Presence*, Notre Dame University Press, Notre Dame, Indiana.

12. Johnson, J.: 1991, 'Historical Roots and Sources of the Just War Tradition in Western Culture', in J. Kelsay and J. Johnson (eds.), *Just War and Jihad*, Greenwood Press, New York, pp. 3–30.

13. Jonsen, A. and Toulmin, S.: 1988, *The Abuse of Casuistry: A History of Moral Reasoning*, University of California Press, Berkeley.

14. Kass, L.: 1985, *Toward a More Natural Science: Biology and Human Affairs*, The Free Press, New York.

15. Langerak, E.: 1994, 'Duties to Others and Covenantal Ethics', in C. Campbell and B. Lustig (eds.), *Duties to Others*, Kluwer Academic Publishers, Dordrecht, The Netherlands, pp. 91–108.

16. May, W.: 1983, *The Physician's Covenant: Images of the Healer in Medical Ethics*, The Westminster Press, Philadelphia.

17. Neuhaus, R.: 1984, *The Naked Public Square: Religion and Democracy in America*, W.B. Eerdmans Publishing Company, Grand Rapids, Michigan.

18. Rojas-Burke, J.: 1994, 'Time to Die: Who Makes the Choice?', *The Register-Guard*, Eugene, Oregon (May 2), 1A, 4A.

19. Toulmin, S.: 1987, 'The National Commission on Human Experimentation: Procedures and Outcomes', in H.T. Engelhardt Jr. and A. Caplan (eds.), *Scientific Controversies: Case Studies in the Resolution and Closure of Disputes in Science and Technology*, Cambridge University Press, Cambridge, pp. 589–613.

20. Toulmin, S.: 1981, 'The Tyranny of Principles', *Hastings Center Report* 11, 31–39.

21. Walters, L.: 1985, 'Religion and the Renaissance of Medical Ethics in the United States: 1965–1975', in E. Shelp (ed.), *Theology and Bioethics: Exploring the Foundations and Frontiers*, D. Reidel Publishing Company, Dordrecht, The Netherlands, pp. 3–16.

22. Weber, M.: 1958, *The Protestant Ethic and the Spirit of Capitalism*, Charles Scribner's Sons, New York.

GERALD R. WINSLOW

Minding Our Language:
Metaphors and Biomedical Ethics

INTRODUCTION

We should watch the way we talk. Human society can be described as a series of long conversations about what matters. In these conversations, the language that we use to describe our social practices not only reveals our values and virtues, it shapes them. Nowhere is this more evident than the language of health care.

I want to consider four ways of talking about health care and ask about their ethical significance. By paying more attention to the way language both stimulates and captivates our moral imaginations, we may be better able to use language that comports with rather than corrupts our moral convictions. In the process of minding the language of health care, we should pay attention to the resources provided by health care's roots in religion as a way of preserving some of health care's most important values.

LANGUAGE AND ETHICS

Philosophers from Plato onward have puzzled over the relationship between language and moral convictions. In the last half of the present century this relationship has been the subject of especially intense study. After the appearance of Wittgenstein's *Philosophical Investigations*, philosophers have devoted countless volumes to the analysis of different modes of expression, or "language-games", and the forms of life which they represent [22]. A common result of such analyses, whether applied to science, or religion, or ethics, has been to conclude that diverse modes of expression reflect different conceptions of reality, each with their own accompanying set of assumptions, relevant actions, and appropriate goals. As Jeffrey Stout remarks in reference to ethics: "The idea that there are distinct moral languages, disparate conceptualities within which to understand and appraise conduct, character, and community, has become a commonplace in recent humanistic scholarship" ([20], p. x).

But such conclusions are not limited to philosophers preoccupied with the

19

E. E. Shelp (ed.), Secular Bioethics in Theological Perspective, 19–30.
© 1996 *Kluwer Academic Publishers. Printed in the Netherlands.*

analysis of language. Anyone who knows more than one language knows how difficult it is to express oneself when the words are missing. For example, I once met an English-speaking pastor whose Alberta congregation included a number of German immigrants. He described a social occasion at which hearty laughter resulted from a joke told in German. When the pastor asked what was funny, the joke teller started to repeat the story in English. Then he hesitated and said, "Taint funny in English!"

The problem is not always just cleverness in translation. The requisite conceptual categories may be absent if the language is missing the needed words. In short, sometimes words really do fail us. An interesting example in biomedical ethics is the expression "informed consent". While on sabbatical at a German university, I was asked by one of my colleagues if I had noticed that there was no standard expression in German for "informed consent". We could think of ways to invent such a German expression. ("Informierte Zustimmung" might have worked.) But any such invention would not have been part of common usage at that time and almost certainly would not have meant what American English had come to understand by informed consent. Thus, it is hardly surprising that German scholars, writing about the concept of informed consent, have sometimes resorted to using the English expression [11].

Lately, I learned that the same situation exists with Japanese. Midst the Japanese characters of a recent newspaper article on the subject of informed consent are the capital letters "IC" which the author uses for the concept of informed consent [23]. It may be that in time "IC" will catch on and take its place in ordinary Japanese usage. Or a comparable Japanese expression may develop. ("Setsumei to Doi", or "explanation and agreement" has been suggested.) If this happens, then the way that the expression is used will not only reflect a form of life but will also influence it.

Within the system of any language, it is the usage of words that determines their meanings. Words have no absolute definitions determined by some celestial lexicographer. They mean what they do when interacting with each other.

Let me illustrate. When my family and I first lived in Austria, my older daughter, then four, began playing with the other children in the park. But there was confusion, rejection, and even a few tears. She could not communicate in German, and insisted that she would never be able to learn how. After only a couple of weeks, however, I observed her one day on the merry-go-round. As she rode, she was shouting to the other children "Schneller! Schneller, bitte!" ("Faster, please!") Later, walking home, I said to her, "I see that you've learned some German". "No", she answered, "I haven't". "But I heard you say 'schneller, bitte" on the merry-go-round", I said, "What does that mean?" She looked puzzled and answered, as only a four-year-old could, "I don't know, Daddy, but it makes the merry-go-round go faster".

As my daughter discovered, words are power tools. They can signal our intentions or desires. They convey our sense of values. And, once established, they can also form our understanding of what is real, what is possible, and what is valuable. "Informed consent" has done this for our culture's form of life. As individuals, we get to contribute very few words to our language. (I doubt that I have added any.) Rather, the powerful tools of language are a gift to us from social interactions within a culture, and as we have seen with "informed consent", sometimes among cultures. Individuals quickly come and go. The language remains, and changes rather slowly, exercising its power for many generations.

THE SPECIAL POWER OF METAPHORS

Among the most powerful tools in language are the figures of speech we call metaphors.[1] They are also among the most difficult to explain in straight forward speech. Consider the difficulty of clarifying for my visiting German cousins sentences such as: "Jim is a square". Or, "In the presence of dogs, most cats are chicken". Even the most thorough dictionary knowledge of the language will not help to make sense of these utterances. Still, it is likely that most native speakers of American English will have little trouble getting the meaning.

Metaphors function by suggesting some resemblance or analogy between two things or experiences that are not literally the same. Metaphors are not, however, mere rhetorical flourishes or linguistic decorations. Rather, they are potent instruments for empowering conceptualization.

The power of metaphors to enliven and shape our moral imagination can be illustrated by examples from health care: "The patient in room 213 is a vegetable". "A fetus is the most common tumor of the uterus". I have heard expressions like these – the first quite often, the second only once. Whether used frequently or seldom, they have common elements. In each case, the metaphor expresses an analogy that highlights some features of reality while obscuring or distorting many others, a point to which I will return to later.

When we hear that the patient is a vegetable, no sensible (or sensitive) person would ask, "What sort of vegetable?" Nor would we be permitted to do to the patient what we typically do to vegetables. We know, or are supposed to know, that the expression does not refer to the patient's skin color or texture but rather refers to the patient's decerebrate state. Calling a patient a "vegetable" fixes this fact in our minds in an indelible way.

But, having disclosed this aspect of reality, the metaphor also obscures a great deal. In most every respect, the patient is not at all like a vegetable. The first time in the history of language that a patient was called a vegetable we may imagine that the expression was surprising and maybe even disturbing. But it has now become so common that it occasions little or no shock. The metaphor has even worked its way into the official language of health care professionals: the "persistent vegetative state".

The second example, the fetus as tumor, is not common. It is jarring.[2] If it were to work its way into ordinary parlance, it would say much about how we evaluate fetal life. If it were to become completely common, it would also develop considerable power to construct the way we or subsequent initiates in our language think about prenatal life.

It is the interaction between metaphorical language and biomedical ethics that I want to explore with four examples from the history of health care. At different times, these four examples have become conventional, metaphorical systems in the everyday language of health care. They are not just single metaphors, but rather systems in which a number of complementary metaphors work synergistically both to describe and to preserve distinct meanings of health care. Moreover, such systems, with their interlocking metaphors, are not easily translated into non-figurative speech. Indeed, it is doubtful that such translation is possible without significant loss of meaning. The following four examples will illustrate why it is important to mind our language.

THE MINISTRY OF HEALING

Probably the oldest way to represent health care in Western culture (and maybe in most human societies) is as religious service. The Oath of Hippocrates, for example, begins with sworn allegiance to the gods: "I swear by Apollo Physician and Hygieia and Panaceia and all the gods and goddesses, making them my witnesses, that I will fulfill according to my ability and judgment this oath and this covenant" ([5], p. 6).

The idea of health care as sacred service was strong not only in ancient Greek culture, but also in the Judeo-Christian tradition. One of the most beautiful expressions of health care as a divinely appointed ministry is the prayer attributed to the famous medieval Jewish physician, Maimonides. It ends with these words: "Thou, All-Bountiful One, hast chosen me to watch over the life and death of Thy creatures. I prepare myself now for my calling. Stand Thou by me in this great task, so that it may prosper. For without Thine aid man prospers not even in the smallest things" ([12], p. 315).

For centuries in the Christian community, the linkage between religious faith and health care was strong. In the Christian testament, one word (*sozo*) means both "to heal" and "to save". The healing and saving ministry of Jesus served as the model for Christians.

Early Christians are known to have established a variety of facilities for the sick and the poor. As early as 330 A.D., St. Helena, mother of Constantine, established a hospital as an act of Christian service. When, late in the fourth century, Emperor Julian the Apostate attempted to return the realm to pagan religion, he blamed Christian hospitals and the women who served in them for Christianity's hold on the common people, and he proposed establishing pagan alternatives as an antidote. By the time of the crusades, most organized health care was delivered by religious orders such as the Order of the Knights of the Hospital of St. John of Jerusalem (or Hospitalers), the Order of the Knights of the Temple of Solomon (or Templars), and the Order of Lazarus which was devoted to the care of lepers. One of these religious orders, San Spirito (or Order of the Holy Ghost), founded hospitals first in Rome and then in nearly all major European cities.

The language of ministry is also found in standard works on medical ethics. For example Percival's book, which held sway in England and America for well over a century, is full of such language. Percival opened his treatise with the words: "Hospital physicians and surgeons should minister to the sick" ([15], p. 71). A few pages later, he wrote: "The physician should be the minister of hope and comfort to the sick. . . . and counteract the depressing influence of those maladies which rob the philosopher of fortitude, and the Christian of consolation" ([15], p. 91). He also suggested that sickness could function to bring people to an openness to religious influence. Physicians should be aware of this opportunity and should also support the work of clergy in their ministry to the sick. Percival even went so far as to insist that hospital rounds and consultations should be scheduled in such a way as not to interfere with religious services.

Other evidence of Western health care's roots in religious faith remain throughout modern cultures, including their languages. In Germany, for example, nurses are still called "krankenschwestern", or sisters for the sick. In England, head nurses are still referred to as "ward sisters". Hospitals named for saints or called Sacred Heart or Good Samaritan are still common. I even see ambulances named "Mercy". Religious communities still own and operate health care facilities throughout the world. Countless thousands of health care professionals worldwide understand their work first of all as serving God by caring for those in need.

The language of health care as a ministry is the language of service, compassion, and covenant loyalty. It is the language of work in a sacred calling.

WAR AGAINST DISEASE

During the second half of the last century a new way of talking about health care arose. Medicine became a war against disease. Susan Sontag suggests that the change in language occurred, in part, because of the rise of germ theory [18]. It is also the case that much of what medicine learned in areas such as trauma surgery and triage was learned on the battlefield. Whatever its source, and however novel it may first have seemed, the metaphor of health care as a battle became commonplace.

In this way of speaking, disease is the enemy, which threatens to invade the body and overwhelm its defenses. Medicine combats disease with batteries of tests and arsenals of drugs. As physicians battle illness, they sometimes refer to their armamentarium. They also write orders. Young staff physicians are still called house officers. At the end of their hospital stay, patients are discharged. Nurses, who take orders, also work at stations. In the past, at least, they also wore uniforms while on duty. But as one early author on nursing decorum wrote, nurses should not wear uniforms when off duty; then, they should wear "civilian dress" ([17], p. 118). As nurses progressed up the ranks, stripes were added to their caps, and insignia pins to their uniforms. Sometimes their orders even called for them to give shots.

The language of military discipline pervaded much of the literature of nursing and medicine early in this century. One nursing leader wrote: "Carrying out the military idea, there are ranks in authority. . . . The military command is couched in no uncertain terms. Clear, explicit directions are given and are received with unquestioning obedience" ([16], p. 452).

The language of medicine as war lives on in many obvious ways. It is still common for politicians to win popular support by calling for a war on some disease such as cancer or AIDS. And in a recent article on medical residencies, the author writes: "[H]ouse officers often relieve the stress of physicians who have completed their training by manning the front lines of acute medical care" ([2], p. 84).

Health care's military language speaks of loyal obedience to authority, and courageous service against a common enemy. To the extent that this language retains its power, it supports a willingness to accept danger, work long hours, and suffer hardships for the sake of winning the struggle against illness.

DEFENSE OF PATIENTS' RIGHTS

In the 1960s and 1970s, another way of talking about health care arose with the patients' rights movement. Sociologist Paul Starr has detailed what he calls the "stunning loss of confidence" sustained by health care professionals in the 1970s ([19], p. 379). Previously, the "sovereign profession" of medicine was largely unchallenged in its authority. But with the rise of consumerism, health care was increasingly depicted as an arrogant and impersonal bureaucracy from which patients, now called clients, deserved protection.

The new metaphorical system drew heavily on legal terminology and traditions. For example, patients would now be protected by a "A Patient's Bill of Rights", authored principally, it might be noted, by an attorney and adopted by the American Hospital Association (AHA) in 1973. In its preface to this document, the AHA states: "The traditional physician-patient relationship takes on a new dimension when care is rendered within an organizational structure. Legal precedent has established that the institution itself also has a responsibility to the patient" ([1], p. 90). The document reassures a patient that she has numerous rights, including the right to full information about her diagnosis, prognosis, and proposed treatments, the right to consent to care, the right to refuse unwanted care, and the right to confidentiality.

The legal nature of this way of speaking was furthered by the language of patient advocacy. Health care professionals, especially nurses, adopted the role of being client advocates.[3] We have only to recall another profession that has clients and refers to its practitioners as advocates to see the legal linkage. Now, instead of being servants of God in the ministry of healing or good soldiers in the war against disease, these new advocates courageously defended the rights of their clients against the overbearing paternalism of earlier tradition. Within this newer metaphoric system, what needs protection, as much as the patient's health, are the patient's rights, especially the right to personal autonomy, as guarded by the practice of informed consent.

The 1970s also saw a dramatic increase in the practice of suing health care professionals, especially physicians, on the grounds that they had failed to honor patients' rights to adequately informed consent. In California, for example, the landmark case of *Cobbs v. Grant* established that a surgeon could be sued successfully not because he was negligent but because "he did not discuss any of the inherent risks of the surgery" with the patient [4].

The legal approach to health care, with its attendant language, is also revealed in the rise of the so-called "advanced directives". Borrowing again from law, patients were given the options of preparing "living wills" or assigning "durable power of attorney for health care". California led the way with both of these approaches, enacting the "Natural Death Act", with its "Directive to Physicians" (California's version of the living will) in 1976 and establishing durable power of attorney specifically for health care in 1983.

Thus, the language that accompanies the defense of patients' rights is a metaphorical system borrowed largely from the law. It emphasizes patient dignity and condemns paternalism. It calls for health care institutions and the professionals who work in them to protect the rights of their clients against any offenders.

HEALTH CARE INDUSTRY

The latest contender for metaphorical dominance comes not from the law but from corporate America. It is the language of the health care industry.

The first person, of whom I am aware, who noticed this new candidate for linguistic dominance was Rashi Fein. He did not like what he was hearing. In a short, 1982 article in the *New England Journal of Medicine*, Fein complained: "A new language is infecting the culture of American medicine. It is the language of the marketplace . . . and of the cost accountant" ([6], pp. 863–864). Fein went on to say that such language is dangerous because it "depersonalizes both patients and physicians".

Despite such protests, the language the health care industry now bids fare to dominate the way we speak of health care. Despite his rather prescient observations, Rashi Fein could hardly have imagined over a decade ago just how pervasive such language would become. It is now so common that one probably runs the risk of sounding like a crank, or worse, for even bothering to notice it. This is the age of managed care, or what some call, managed competition. Consider the following now ordinary expressions, all taken from one article on health care management: "Total quality management" is a system intended to "reduce costs" while maintaining quality in "a highly competitive market". Today's health care must "identify its customers" and understand the "business connotation of customer-supplier". "Competitive advantage improves as a result of improved quality and lower costs" ([13], pp. 274–275).

With increasing frequency, one now hears hospitals refer to their product line, their market share, their human resources, and even their guests. The industry is increasingly customer driven, cost conscious, and productivity oriented. Treatments are more and more evaluated for their cost worthiness by using complicated cost-benefit ratios. Health care professionals are now providers who give their customers (or guests, or consumers) what the customers want and at the lowest possible price. One revealing sentence in a recent edition of *Medical Economics* says it all: "In management you'll have to switch your concern from the well-being of the patient to the health of the bottom line"([14], p. 166).

The language of the health care industry is borrowed from economics and business. It has the feel of hard-edged, economic realism. Health care is depicted as a business in which powerful enterprises vie with their competitors. The language of industry makes it seem normal to attend to marketing, competition, and customer satisfaction.

METAPHORS AND MORAL VALUES

How do metaphoric systems, like the four just mentioned, interact with our moral values, judgments, and actions? I want to suggest four features that deserve our attention.

1. Metaphors are powerful partly because of their capacity to highlight certain aspects of reality in memorable ways while obscuring other aspects of reality.

We do well to pay attention not only to what the metaphorical systems, which I have outlined, disclose but also what they may obscure or distort. To call health care a ministry is to emphasize faithful service, devotion, and compassion. In all honesty, however, we must also say that the language of the ministry of healing has sometimes obscured the fact that such care requires a sound scientific basis and has to be paid for in some way. Even the most charitable ministries of healing were also economic realities. Failure to notice the fact that ministers of healing must also be adequately compensated for their services could easily lead to the exploitation of some health care professionals. Ministering beneficently to the needs of others may also lead to attitudes and actions of paternalism, in which the patient's own preferences are given little heed.

Similarly, health care as war highlights the struggle against a common enemy. This metaphor can awaken courage and a willingness to sacrifice for a good cause. But the metaphor has also worked very effectively to preserve patterns of authoritarian relationships, with unquestioning obedience and loyalty, that few would find healthy today.

The legal metaphor of patients' rights focuses on the inviolability of each person. But an overemphasis on honoring rights may obscure patients' responsibilities for their own health. The legal language may also distort the reality of intimate human relationships within families and between families and their professional care givers.

The industrial metaphor reminds us that health care is, and always has been, an economic enterprise. Still, an emphasis on competitive advantage in seeking customers for health care's products may obscure or miserably distort the virtues necessary to care genuinely for sick people.

2. Metaphorical systems make some actions seem normal and expected, while making other actions seem strange or unacceptable.

Once established, a way of speaking about health care can affect attitudes which, in turn, will affect behavior. If, for example, the military metaphor is dominant, normal actions would include obeying orders, respecting one's superiors, wearing uniforms, taking risks (including the risk of death), feeling and being loyal, and suffering hardships such as night duty and low pay. At the same time, other actions would seem strangely unacceptable: questioning orders or disobeying them, refusing to accept risks, sharing secrets, and demanding higher pay.

While they are dominant, metaphorical systems often work their way into the structure of social institutions at both formal and informal levels. It is not uncommon for such metaphors even to become part of social policy and law. The legal metaphor of being a patient advocate was once a novel idea. Now it is part of California's law. The state's Nurse Practice Act now includes in the list of standards of competence the following: "A registered nurse shall be considered competent when he/she . . . acts as the client's advocate, as circumstances require, by initiating action to improve health care or to change decisions or activities which are against the interests or wishes of the client, and by giving the opportunity to make informed decisions about health care before it is provided" [3].

One fine institution, Sarah Lawrence University, even offers a masters degree in Health Advocacy. And *Index Medicus* now lists "patient advocacy" as a separate category. A computer search of one recent year's entries indicated that there were 167 articles under this heading – not bad for a locution that was new-fangled just two decades earlier.

3. Metaphorical systems become most powerful when their usage is least noticed.

The power of such language to focus our attention and grab our imagination grows as it becomes time-worn. This is so because the ability to highlight some aspects of reality, while obscuring others, becomes greater when common usage has caused the metaphorical nature of the language to be lost from view. It is just when such language becomes unremarkable that it has the power to make some character traits appear virtuous and some vicious, some actions normal and expected and others simply strange. So, for example, the jargon of the health care industry, which occasioned the ire of Rashi Fein a dozen years ago, is now so common that it goes almost entirely unnoticed. To attend the hospital's marketing committee and discuss its product line or market share is now normal. It would require a tenacious memory to recall the time when such language was new.

4. Metaphorical systems are dominant only for a time.

In time, they fade and find their hegemony displaced by a new system that has slipped into our ordinary speech, generally unannounced. Rarely, if ever, are metaphorical systems destroyed by design, because critics found them unworthy. Nor is it that the older ways of speaking grow too trite. It is that they cease to be effective. They fade into the background of usage, gradually becoming impotent or unheard. Such metaphorical systems, just like many Native American languages, become lost, sometimes irretrievably. In this regard, language is like a giant attic with discarded items from an earlier time gathering dust, largely, if not entirely, forgotten. So, today, one does not hear much about the ministry of healing. Nor will it be found in *Index Medicus*. Perhaps, in the future, we may imagine some scholar with a grant from the National Endowment for the Humanities going around with a tape recorder looking for health care professionals whose mother tongue was the language of ministry, and who can still remember a few phrases.

CONCLUSION

It is a long way from the ministry of healing to the health care industry. The distance is best measured not in miles or years but as the distance between sets of values and virtues – the forms of life that these ways of speaking represent and inform. It is, of course, still possible in our culture to pull the language of ministry out of the attic, dust it off, and use it again. Despite what our high school composition teachers taught us, metaphors can even be mixed. Consider the expression, "No margin, no mission". I have sometimes heard this around health care institutions with religious heritages. But within the context of a pluralistic, secular culture, we may expect to hear such language less frequently.

Does this mean that the earlier languages of health care, such as the ministry of healing, are bound to be lost? I think not. But their preservation may require more explicit intentionality, especially as the language of the market becomes further entrenched and thus further hidden from view. It may, for example, be part of the mission of religiously based health care institutions to preserve, by design, the language of health care as a ministry. As with the other ways of speaking that I have identified, loss of the language of ministry could represent a loss of significant values in health care.

It is, of course, not the language itself that is finally important. It is the form of life to which the language bears witness that matters. One of the special contributions of communities of memory, including religious communities, is the preservation of the modes of expression that shape and sustain important values. People of religious faith have a distinctive task in this regard. They cannot expect others to speak their language for them. Nor should they imagine that adhering to their mission is easy. To understand this, we have only to think again of all the native languages that have become extinct in our society, even when heroic measures were taken to preserve them. For members of a culture in which the language of the market now dominates, it will be increasingly difficult to remember that health care can be, first of all, a spiritual calling.

In order to understand the importance of what is at stake, we have merely to consider what would be lost if the industrial metaphor were to become the only language used to describe the meaning of health care. As Jeffrey Stout observed in another context: "Some languages, in particular those of the marketplace and the bureaucracies, creep into areas of life where they can only do harm. They tend to engulf or corrupt habits of thought and patterns of interaction that we desperately need" ([20], p. 7). The fundamental tasks of health care still require sincere, courageous compassion in the face of illness and eventual death. The language of spiritual ministry has supported such care in powerful ways. The preservation of such language is not impossible. And it is worth the trouble.

I have not argued that the metaphors of war, law, and business have no useful place in health care. Clearly, they do. What is equally clear, I think, is that the metaphors of ministry should not be lost. They can still provide powerful reminders of the central purpose of health care, to serve those who are in need. Finding fresh ways to keep the language of ministry alive should be a welcome task for those of us who find in health care an opportunity for spiritual service.

I conclude with a story that, I trust, will explain itself. Not long ago two of my favorite uncles, whose first language was German, came from the mid-West to visit my mother in Oregon. I made a special trip there to see them again and to video tape these octogenarians as they told stories from the past. After awhile, I suggested that we use the speaker phone to call one of our cousins in Germany, one about their age. I read embarrassment on both of their faces. Then, with reticence, they explained that they had not spoken German in many years and that there was no way they could carry on a conversation. Never mind, I said, I would translate for them. I made the call. Then something interesting happened. About one minute into the conversation, first one and then the other uncle began to join the conversation without my help. The pace picked up. A certain joy was obvious, and I was, for the most part, sidelined as three cousins swapped stories in a language that two of them were sure had been lost. Finally, I had to suggest that it was time to say "aufwiederhören".

NOTES

1. My thinking about metaphors is influenced in a number of ways by the work of George Lakoff and Mark Johnson ([8], [9], [10]).
2. A physician colleague informs me that the original and still primary medical meaning of "tumor" is simply "swelling" (from the Latin *tumere*, "to swell"), a fact confirmed by my medical dictionary ([7], p. 1660). So the use of "tumor" to refer to pregnancy is not as unusual as it may initially seem.
3. The development of the military metaphor is elaborated in greater detail in an essay by Gerald Winslow [21].

BIBLIOGRAPHY

1. American Hospital Association: 1989, 'A Patient's Bill of Rights', in R.M. Veatch (ed.), *Cross Cultural Perspectives in Medical Ethics*, Jones and Bartlett, New York.
2. Aronowitz, R.A.: 1990, 'Residency as Metaphor', *Journal of General Internal Medicine* 5, 84.
3. California State Nursing Practice Act: 1992, paragraph 1443.5.
4. *Cobbs v. Grant*, 502 P.2d 1 (October 27, 1972).
5. Edelstein, L.: 1989, 'The Hippocratic Oath: Text, Translation, and Interpretation', in R.M. Veatch (ed.), *Cross Cultural Perspectives in Medical Ethics*, Jones and Bartlett, Boston.
6. Fein, R.: 1982, 'What is Wrong with the Language of Medicine?', *The New England Journal of Medicine* 306, 863–864.
7. Friel, J.P.: 1974, *Dorland's Illustrated Medical Dictionary, 25th ed.*, W.B. Saunders, Philadelphia.
8. Johnson, M.: 1993, *Moral Imagination*, The University of Chicago Press, Chicago.
9. Lakoff, M. and Johnson M.: 1980, 'Conceptual Metaphor in Everyday Language', *The Journal of Philosophy* 78, 453–486.
10. Lakoff, M. and Johnson, M.: 1980, *Metaphors We Live By*, University of Chicago Press, Chicago.
11. Laufs, A.: 1990, 'Fall und Commentare: Therapie beschrankung durch Patienten willen?', *Ethik in der Medizin* 2, 90–93.
12. Lyons, A.S. and Petrucelli, R.J.: 1978, *Medicine: An Illustrated History*, Abradale Press, New York.
13. Masters, M. and Masters, R.: 1993, 'Building TQM into Nursing Management', *Nursing Economics* 11, 274–275.
14. *Medical Economics*: 1993 (unsigned editorial comment), 70, 166.
15. Percival, T.: 1803, *Medical Ethics*, S. Russell, London.
16. Perry, C.M.: 1906, 'Nursing Ethics and Ettiquette', *The American Journal of Nursing* 6, 452.
17. Robb, I.H.: 1900, *Nursing Ethics*, E.C. Koeckert, Cleveland.
18. Sontag, S.: 1978, *Illness as Metaphor*, Farrar, Straus and Giroux, New York.
19. Starr, P.: 1982, *The Social Transformation of American Medicine*, Basic Books, New York.
20. Stout, J.: 1988, *Ethics after Babel*, Beacon Press, Boston.
21. Winslow, G.R.: 1984, 'From Loyalty to Advocacy', *The Hastings Center Report* 14, 32–40.
22. Wittgenstein, L.: 1953, *Philosophical Investigations*, trans. by G.E.M. Anscombe, Macmillan, New York.
23. Yamagata, K.: 1994, 'Essay: Informed Consent', *Kobe Times*, January 19.

B. ANDREW LUSTIG

Reform and Rationing: Reflections on Health Care in Light of Catholic Social Teaching[1]

INTRODUCTION

In this essay, I analyze two issues at the forefront of recent policy discussion – health care reform and health care rationing – in light of Roman Catholic social teaching. It is important, as I begin, to identify and critique a position often deemed to be the conventional wisdom on the place of religion in policy debates. Many would assert that the claims of particular religious communities are irrelevant or unnecessary to the formulation and justification of public policy in a secular and pluralistic society. However, their easy dismissal of religious voices can be questioned on two grounds. First, they assume, with little or no argument, that the claims of religious communities cannot be justified in more general terms. If such common ground were indeed absent, it would seemingly follow that religious communities could not fully participate in policy formation because of constitutional concerns regarding the separation of church and state. However, this first assumption is demonstrably false. The facts of overlapping consensus, despite the varied theological, philosophical, and political viewpoints of those who achieve practical agreement, play a pivotal role in legislative pronouncements and judicial reasoning. Shared moral and legal conclusions on particular issues and common policy conclusions on appropriate societal remedies are possible despite the lack of agreement about the first principles at work in particular perspectives, whether religiously informed or not. Second, the tendency to disqualify the claims of religious communities from debates on public policy confuses the *process* of public discussion and debate, where the views of particular communities may exercise legitimate influence, with the *warrants* for and *justification* of policy choices, where parochial appeals are inappropriate.

The present essay is offered as an exercise meant to illustrate, through the lens of debates on health care reform and rationing, how Roman Catholicism has engaged the problematic of particularism in public debate and discussion. The essay falls naturally into three parts. After this brief introduction, I consider, in Part I, the theological warrants in modern Catholic social teaching on the nature and scope of health care as a positive right. In

31

E. E. Shelp (ed.), Secular Bioethics in Theological Perspective, 31–50.
© 1996 *Kluwer Academic Publishers. Printed in the Netherlands.*

Part II, I consider the legitimacy of health care rationing in light of Catholic social principles that constrain health care as an individual entitlement according to the requirements of the common good. In Part III, I assess the relevance of Catholic arguments on health care reform and rationing to the broader secular debate on those issues. I conclude by suggesting that the Roman Catholic approach, while sharing affinities with consensus principles that can be developed on other grounds, also brings a distinctive perspective to issues of health care policy.

I. Health Care in Catholic Social Teaching

In Catholic social teaching since Vatican II, access to health care is seen as a positive right, i.e., a justified entitlement claimable by individuals against society. The warrants for that position are expressly theological, involving a number of themes and principles that, while interconnected, can be analyzed separately. The first is an appeal to the dignity of the individual made in the image of God. The second is an understanding of the common good, which in contrast to secular liberal understandings, presents an organic vision of society with duties incumbent on institutions according to the purposes of society as established by God. The third theme, which is developed in the modern encyclical literature as an extension of the traditional emphasis upon the common good, is the regulative ideal of social justice. Social justice is a specific notion that enjoins institutions and, increasingly, governments to guarantee the basic material concomitants of individual dignity. The fourth theme involves an appeal to the principle of subsidiarity, first enunciated by Pius XI, which speaks to the intrinsic and instrumental value of meeting the basic needs of persons at the lowest or least centralized level of association and authority possible. Finally, certain Catholic writers have emphasized distributive justice as a decisive appeal that functions, in important respects, independently of the general institutional focus expressed in the language of social justice. Each of these themes or principles merits careful attention.

In Catholic social teaching, every individual has dignity because he or she is made in the image and likeness of God (Genesis 1:26) and has been redeemed by Christ (Ephesians 1:10). By the time of John XXIII, the material conditions required by individual dignity have been enumerated into a list of specific individual rights, including the right to health care, which must be safeguarded by responsible institutions. John says in *Pacem in Terris*:

> . . . we see that every man has the right to life, to bodily integrity, and to the means which are necessary and suitable for the proper development of life. These means are primarily food, clothing, shelter, rest, *medical care*, and finally the necessary social services [emphasis mine]. Therefore, a human being also has the right to security in cases of sickness, inability to work, widowhood, old age, unemployment, or in any other case in

which he is deprived of the means of subsistence through no fault of his own ([5], p. 167).

In John's successor, Paul VI, one finds a renewed emphasis on a theme central to the writings of Pius XI, namely, that "material well-being is not simply instrumental in value. It is not a means of a dignified life. It is, rather, *integral* to the standard of all moral value, human dignity" ([8], p. 79).

Throughout this discussion, one is struck by the general nature of papal pronouncements about what individual dignity requires. That level of generality is not surprising. The encyclicals are statements of theological and moral vision rather than policy recommendations. They provide more than a mere statement of ideals but less than a blueprint for specific choice and action. Nonetheless, the expressly theological basis of human dignity provides, albeit in quite general terms, a different and richer context for understanding the usual arguments about liberty and equality in secular debates about the right to medical care. In Catholic social teaching, although individuals are endowed with freedom, theirs is a freedom to be exercised in community. The latter emphasis is surely to be expected, given the trinitarian nature of the God whom Christians worship and in whose image they are made. That trinitarian understanding has specific moral implications. In the words of one commentator:

> . . . our trinitarian theology becomes a radical challenge to community. How can Christians say they believe in God if they are unwilling to put together structures that build human community and meet fundamental human needs? How can someone really claim to believe in the triune God and not feel a sense of outrage about the quarter of the U.S. population which lacks or is inadequately supplied with such a basic good as health care coverage? If we believe in the triune God as the very ground of community, the problem of our health care system is not just an ethical or economic or political problem. The problem is ultimately a religious or theological problem ([10], p. 105).

In light of the belief in God as Trinity, and in contrast to the polarities one often finds in secular debates about positive rights, Catholic social teaching elevates neither liberty nor equality to a position of unchallenged priority. Rather than seeing either value as trumping the other, Catholic teaching, especially since the time of John XXIII, presents liberty and equality as mutually accommodating principles.

The notion of the common good is a second characteristic emphasis in recent Catholic social teaching. Again, this appeal, much like the theological grounding for arguments about individual dignity, offers an alternative to those secular versions of political theory that emphasize liberty as either a side-constraint or as a primary positive value. Unlike approaches that begin (and often end) with an emphasis upon individualism, the common good is fundamentally social and institutional in its focus. It stresses human dependence and interdependence. However, the common good should not be interpreted as a Roman Catholic analogue to a utilitarian calculus. Rather, the common good is the "set of social conditions which facilitate the realization of personal goods by individuals" ([8], p. 64). The common good

> insists on the conditions and institutions . . . necessary for human cooper-
> ation and the achievement of shared objectives as decisive normative
> elements in the social situation, elements which individualism is both unable
> to account for in theory and likely to neglect in practice ([11], p. 102).

The common good need not be viewed as being in necessary tension with the rights of individuals, at least in encyclicals since the time of John XXIII. Rather, the common good functions in two important senses: first, it is invoked to temper and correct the inequities often associated with secular individualism; and second, it incorporates guarantees for personal rights and duties ([12], pp. 429–436). Still, in contrast to individualistic theories, a fundamentally social understanding infuses Catholic social thought. This social perspective is especially evident in Catholic teaching on the nature and scope of property, with implications for a number of positive rights, including the entitlement to health care. Theologically, men and women are imagers of a trinitarian God. Practically, this suggests that the claims of individuals to resources are limited by the claims of others for the satisfaction of basic needs. In hard cases, where choice is inevitable, John Paul II characterizes the conclusions of the tradition as follows:

> Christian tradition has never upheld [private property] as absolute or
> untouchable. On the contrary, it has always understood this right within
> the broader context of the right common to all to use the goods of the whole
> of creation: the right to private property is subordinated to the right to
> common use, to the fact that goods are meant for everyone ([9], # 43).

In this emphasis on property in common as a regulative notion, the common good emerges as fundamental. Individual rights and duties are seen as constitutive of the common good, but there are no absolute or unmediated claims to private ownership of property. Unlike perspectives that begin with the distinction between private and public resources as the unassailable datum from which moral analysis proceeds, the Catholic tradition does not deem individual ownership to be sacrosanct. Rather, common access according to use remains the relevant criterion according to which social arrangements and practices must be assessed, especially in the circumstances of a developed economy.

As a third theme, the Catholic tradition has emphasized, since its enunciation by Pius XI in *Quadragesimo Anno*, the so-called principle of subsidiarity. In one respect, this principle can be seen as following logically from the idea of the common good ([1], p. 132). Drew Christiansen describes subsidiarity as "another dimension" of the common good; subsidiarity, then, involves "the notion that responsibility is rightly exercised at the smallest appropriate level" ([3], pp. 46–47). Thus, problems that can be addressed by the small-scale initiatives of individuals or voluntary assocations should be handled at those levels. Matters that can be solved by lower levels of governmental involvement (city rather than state, state rather than federal) should also be addressed at the lowest level possible to achieve effective results. As a functional corollary of the common good, such a prudential concern with efficiency is quite appropriate. Subsidiarity implies that "the first responsibility in meeting human needs rests with the free and competent individual, then with the local group" ([1], p. 132). Moreover, there is essential, not only instrumental, value in the moral involvement of individuals in the interpersonal forms of association that subsidiarity commends as a functional aspect of the common good. The common good, as I noted, involves the good of persons. Subsidiarity, as an expression of the common good, involves the intrinsic value of direct and immediate forms of the individual's responsibility to others.

However, since the papacy of John XXIII, there has been an increasing emphasis in the Church's social teaching on the necessity of governmental involvement in meeting the basic needs of persons. The principle of subsidiarity, then, bears two implications for health care reform proposals. As Keane observes:

> . . . Catholic social teaching on subsidiarity . . . can tell us two important things. First, we really do need to find a health care system which offers us some balance between private management and public management . . . Second, in view of the increasingly complex character of health care delivery, it is only to be expected (and clearly "catholic") that more and more government management of health care will in fact be necessary. Thus we ought not to be afraid on religious/moral grounds of the fact that many of the major health care reform proposals which are being discussed in the United States today are calling for a significantly increased level of government involvement, at least in the financing of health care. In the end, Catholic social teaching may well help us to decide which of the current reform proposals seem best, by challenging us to consider carefully which proposal best mixes subsidiarity and socialization in our current health care context ([10], pp. 149–150).

To be sure, given the complex features of modern social and economic life, the links between property in common and the common good cannot be understood literally. Thus, since the time of Pius XI, the common good is usually linked with another idea characteristic of recent social teaching, viz. social justice. Pius XI invokes the notion of social justice as a "conceptual tool by which moral reasoning takes into account the fact that relationships between persons have an institutional or structural dimension" ([8], p. 54). In contrast to atomistic understandings of individual rights, Pius emphasizes social justice as a regulative institutional principle. As societies develop, as medicine progresses, institutions, especially at the governmental level, are morally required to mediate the claims of human dignity and to shape the content of human rights, including the right to medical care.

Another emphasis in recent Catholic social teaching has been the so-called "preferential option for the poor". Although this option is sometimes invoked as a separate appeal, it can be viewed as a practical implication of the three broader themes discussed above: the positive rights of individuals, the common good, and social justice. In service to these values, institutions are required to respond to those inequities between and among individuals that particularly threaten the dignity of the most disadvantaged in society. Whether one views the preferential option for the poor as an implication of more fundamental values, or as having independent standing, the practical results at the level of social policy, about health care and other social goods, are likely to be the same.

Finally, some recent perspectives have emphasized the need to invoke distributive justice as an appeal that functions somewhat independently of the more general focus of social justice. Philip Keane is representative of this trend, especially in relation to arguments about health care reform. Although Keane is sympathetic to the broad critique of health care institutions afforded by the language of social justice, he is skeptical about the precision it affords regarding the practical questions associated with health care reform and rationing. Thus he says:

> . . . reform of the structures of society so that society can more effectively deliver health care still depends on society's having a clearer focus on just what health care goods it ought to be delivering to people. Thus, while the social structures question, like the equality question, is a pivotal aspect of all justice including health care justice, my judgment is that the distribution question (i.e. What health care benefits must we provide?) is the most central of all the justice questions which relate to health care ([10], p. 138).

Keane continues in a cautious vein:

> If we focus too much on social justice, the risk is that we will emphasize the structures necessary to furnish health care instead of first focusing on the human need of real persons to have real health care crises adequately addressed. Such emphasis on structures without a prior commitment to genuine human needs can raise all the fears of complex bureaucracies without really substantial goals ([10], p. 139).

For these reasons, Keane believes that the principle of distributive justice should assume moral priority in debates about reform and rationing. In developing this claim, he appeals to the work of Protestant theologian Gene Outka on the requirements of distributive justice in health care [13]. Outka views health care needs as discontinuous from other basic needs. According to Outka, medical needs are randomly distributed, to significant degree "unmerited", often catastrophic, and to a great extent unpredictable. Outka therefore sees critical health care emergencies as having an immediacy and urgency that distinguishes them from other sorts of deprivation. Given that distinctiveness, Outka concludes that "similar treatment for similar cases", based on a criterion of medical need, is the fairest canon of distributive justice among various alternatives that have been proposed for health care delivery. Keane characterizes Outka's conclusions in this way:

> . . . with careful reflection and dialogue, we should determine those health care needs which we as a society must meet as a minimum standard, and then provide those needs for everyone, regardless of merit, usefulness, economic ability, etc. Such an approach means that there will be some possible health services which will not be provided because they are not truly needs or because they are beyond our capacity as a society of mortals who must face the fact of death ([10], p. 142).

II. Two Recent Catholic Documents on Health Care

I now turn my attention to two recent Catholic documents that focus specifically on health care, one the U.S. Bishops' "Resolution on Health Care Reform", the other a moral analysis of health care rationing by the Catholic Hospital Association entitled *With Justice For All?* I have several reasons for analyzing these documents in some detail: first, to discuss the warrants at work in their arguments; second, to scrutinize and critique several conceptual and practical tensions in their respective recommendations for public policy; and finally, to use the documents as points of reference for Part III's brief overview of the relevance of theologically based arguments to secular policy debates.

A. *The U.S. Bishops' Resolution on Health Care Reform*

The recent "Resolution on Health Care Reform" by the U.S. Roman Catholic bishops provides the occasion for a closer look at the way that basic principles of Catholic social teaching on the right to health care have been brought to bear on general issues of health care reform. In its introduction, the resolution identifies the fundamental problems that beset present health-care delivery: excessive costs, lack of access for many Americans, and questions about the quality of the care provided. The bishops address the document to the "Catholic community" *and* to "the leaders of our nation" ([15], p. 98). The document, although at times expressly theological, also attempts to speak to the broader public, primarily through its appeals to human dignity, which can be justified for both theological and non-theological reasons, and by its expressions of concern about the plight of those presently underserved by the U.S. health care system.

The resolution appeals to the recent tradition by rooting its approach to health care in the three fundamental themes I discussed in Part I. First, "[e]very person has a right to adequate health care. This right flows from the sanctity of human life and the dignity that belongs to all human persons, who are made in the image of God". Health care is a "basic human right, an essential safeguard of human life and dignity". Moreover, the bishops' call for reform is rooted in "the priorities of social justice and the principle of the common good". In light of these fundamental values, the bishops judge that "existing patterns of health care in the United States do not meet the minimal standard of social justice and the common good". Indeed, "the current health care system is so inequitable, and the disparities between rich and poor and those with access and those without are so great, that it is clearly unjust" ([15], p. 99).

The bishops also appeal to a preferential option for the poor as a particular implication of the common good. Pointing out that the "burdens of the system are not shared equally", the bishops conclude that we must "measure our health system in terms of how it affects the weak and disadvantaged". Fundamental reform must be especially concerned with "the impact of national health policies on the poor and the vulnerable". In this context, the bishops quote with approval a recent ecumenical statement on the common good; to wit:

> More than anything else, the call to the common good is a reminder that we are one human family, whatever our differences of race, gender, ethnicity, or economic status. In our vision of the common good, a crucial moral test is how the weakest are faring. We give special priority to the poor and vulnerable since those with the greatest needs and burdens have first claim on our common efforts. In protecting the lives and promoting the dignity of the poor and vulnerable, we strengthen all of society ([15], p. 98).

As a final appeal in Section I of the Resolution, the bishops invoke the prudential notion of "stewardship". The cost of present health care in the United States "strains the private economy and leaves too few resources for housing, education, and other economic and social needs". In response, "[s]tewardship demands that we address the duplication, waste and other factors that make our system so expensive" ([15], p. 99).

The practical focus of Section Two of the bishops' resolution raises the same cluster of thematic concerns in more focused fashion. In this section, the bishops set forth eight practical criteria by which to judge the moral adequacy of proposals for reform. The criteria are as follows: (1) respect for life, (2) priority concern for the poor, (3) universal access, (4) comprehensive benefits, (5) pluralism, (6) quality, (7) cost containment and controls, and (8) equitable financing. Given the brevity of the resolution, these criteria are not adequately developed, but as a whole, they emerge as practical expressions of the fundamental theological values that inform the Catholic tradition on health care. At the same time, because the bishops clearly intend these criteria as useful guides for assessing public policy, it is important to note certain conceptual and practical tensions between and among the various criteria. For purposes of the present discussion, I will focus on the first four.

Overall, the criteria are based on considerations of human dignity that are theologically based: thus, "reform of the health care system which is truly fundamental and enduring must be rooted in values which reflect the essential dignity of each person, ensure that basic human rights are protected, and recognize the unique needs and claims of the poor" ([15], p. 100).

The first principle, respect for life, speaks to the need for any reform proposal to "preserve and enhance the sanctity and dignity of human life from conception to natural death" ([15], p. 100). Nonetheless, "sanctity of life" does not, of itself, shed light upon *how* to proceed in cases when "hard choices" involving allocation of resources must be made between and among individuals, all of whom might at least marginally benefit from continued provision of care. Moreover, the relevance of this first principle to certain "hard cases" (including abortion and the status of persistently vegetative patients) is unclear, since the grounding of "personhood" claims in these instances may depend on theological understandings of "ensoulment" or capacities to "image" God that are not available as warrants for secular policy.

The second principle – priority concern for the poor – implies, according to the resolution, that any reform proposal should be judged as to "whether it gives special priority to meeting the most pressing health care needs of the poor and underserved, ensuring that they receive quality health services" ([15], p. 100). Here the difference between according the "preferential option for the poor" independent weight or interpreting it as a particular implication of the common good may significantly affect how one assesses a particular reform proposal. While it is doubtless true that working correlations between overall indices of poverty and poorer health can be drawn, it is not the case that for any particular indigent individual, health outcomes will necessarily be correlated with access to health care. Indeed, there are strong arguments, doubtless of a more prophetic sort, that reform of health care as a discrete sector may be one of the least effective ways of improving the *general* health outcomes of the poor as a class. Overall poverty, not simply limited access to medical care, may be a far more relevant determinant of health than the bishops' second criterion suggests.

With regard to medical care, then, the preferential option for the poor will require greater attention to a definition of the "poor" for whom preference is to be shown – those who are "generally poor", according to most indices, or those who are "medically indigent". To the extent that the unclarity persists, the relevance of the criterion for any particular patient, as compared with its relevance as a working generalization about classes of persons, will not be obvious. Indeed, for any particular patient, the first criterion, "respect for life", might suggest that general indices of poverty, in contrast to criteria relating specifically to medical need or medical indigence, may discriminate unfairly against a needy patient who would not (at least initially) qualify according to the former indices.

By contrast, if the "preferential option for the poor" does not have independent weight but instead is seen as a particular application or implication of the common good, then determinations of medical indigence or medical need might proceed apace with difficult judgments about basic social goods other than medical care. On this account of the preferential option, the common good might be invoked as a systematic consideration to limit the availability of resources for particular individuals as the result of prior social choices, independent of individual circumstances of need. In this scenario, one might then be able to distinguish circumstances that are admittedly unfortunate from those that are unfair ([4], pp. 342 ff.).

According to the bishops' third criterion, any morally acceptable proposal must provide "universal access to comprehensive health care for every person living in the United States". A number of practical questions arise here. How shall the system be reformed in order to provide genuinely universal access to comprehensive benefits? What implications might this principle have for conscripted medical service to presently underserved areas, especially in rural America? Alternatively, if one relies on market mechanisms and incentives to expand access and availability, how will such competitive economics work to ensure entitlement on a universal basis? In addition, should citizenship be morally decisive in determining access to available services? The wording of this criterion would suggest that health care coverage be limited to those within the United States but also that all persons within U.S. borders (not only citizens) be provided care. Practically, this may pose significant difficulties, since the costs of such care, indiscriminately provided, may undercut the willingness and/or ability of taxpayers to fund a basic level of health care for all.

The fourth criterion offers a benchmark of "comprehensive benefits" for any morally acceptable reform proposal. Comprehensive benefits include those "sufficient to maintain and promote good health, to provide preventive care, to treat disease, injury, and disability appropriately and to care for persons who are chronically ill or dying" ([15], p. 100). Again, the choices and tradeoffs required among these various goods might well conflict with the focus on particular individuals seemingly implied in such principles as "respect for life" or "universal access". Health care "sufficient to maintain and promote good health" for a given individual may be, in effect, a black hole, since that person's needs, according to the fourth criterion, might swallow an inordinate amount of resources to which others with lesser needs might otherwise have access. Moreover, as noted above, medical services may be fairly low among overall indices for health. Thus criteria for "health care" reform, if invoked without sufficient attention to the multifactorial nature of good health as an outcome, may be coopted by tendencies already present in technology-driven medicine to misallocate funds that could be better spent on primary or preventive care. Finally, only "caring for" those who are chronically ill or dying might involve a number of rationing choices that, while supported on grounds of the common good or social justice, fail to comport with the individual focus of other criteria.

As noted above, Section Two of the resolution sets forth a number of other practical criteria for assessing reform proposals, which, while interesting, emerge as fairly "commonsensical" in tone. By contrast, Section Three lists four "essential priorities" that the bishops urge upon their readers in applying their eight criteria of assessment: (1) priority concern for the poor and universal access; (2) respect for human life and human dignity; (3) pursuing the common good and preserving pluralism; and (4) restraining costs ([15], pp. 100–101). Although each of these priorities merits scrutiny in its own right, I will focus on key tensions generated by the bishops's discussion of the first and third priorities.

In their discussion of the first priority – concern for the poor and universal access – the bishops voice strong support for "measures to ensure true universal access and rapid steps to improve the health care of the poor and underserved" ([15], p. 100). In light of that commitment, they "do not support a two-tiered health system since separate health care coverage for the poor usually results in poor health care. Linking the health care of poor and working class families to the health care of those with greater resources is probably the best assurance of comprehensive benefits and quality care" ([15], pp. 100–101). Nonetheless, in discussing their third priority – preserving pluralism – the bishops emphasize the following:

> We believe the debate can be advanced by a continuing focus on the common good and a healthy respect for genuine pluralism. A reformed system must encourage the creative and renewed involvement of both the public and private sectors. . . . It must also respect the religious and ethical values of both individuals and institutions involved in the health care system ([15], p. 101).

While it is true that both priorities, as normative generalizations, may help frame policy discussions on health care reform, the potential for tensions between a commitment to a single-tiered system and a respect for pluralism of individuals and institutions is, as a matter of practical policy choice, enormous. Moreover, in light of the theological convictions central to the broader Catholic discussion of health care, considerations of individual liberty and dignity, as well as of the common good, might reasonably lead to a different practical conclusion; viz., that two tiers of health care delivery are both morally appropriate and practically preferable, so long as universal access to comprehensive basic care is assured. Consider, for example, basic education as a useful analogy to health care: a tax-based commitment to education for all citizens does not prevent individual parents from paying for alternative "basic" schooling for their children. It is not obvious, at least without a great deal more argument than the bishops provide, that health care should be viewed differently. Nor is it obvious, in light of the general theological warrants analyzed in Part I of this essay, that the common good would necessarily dictate the priority of equality over liberty in the delivery of health care.

The bishops' tendency to ignore or underemphasize conceptual and practical tensions among criteria and normative priorities is instructive for several reasons. First, it indicates that the bishops often conflate hortatory and pragmatic concerns in their discussion. Second, it suggests that a great deal more by way of careful, practically oriented, argument will be necessary before their assessment criteria and normative priorities can be viewed as significant contributions to the policy debate. Third, as I will conclude in Part III, the conceptual and practical tensions in the bishops' discussion will require the discriminating reader to identify the various modes of moral discourse at work in their recommendations in order to appreciate the different ways that Catholic themes and principles may be relevant to secular discussions.

B. *The Ethics of Health Care Rationing*

In Part I we discussed a number of relevant theologically grounded principles in Catholic social teaching on health care. We have seen how the bishops' resolution addresses general issues of health care reform. I turn now to a consideration of how Catholic themes and principles may function in judgments about the legitimacy of health care rationing. Here, because I am considering rationing along the lines developed by the Health Services Commission in Oregon, a brief review of the Oregon plan is in order.

1. *The Oregon Model of Health Care Rationing*

In 1989, the Oregon State legislature passed the Oregon Basic Health Services Act. The Act established a Health Services Commission, which was charged to develop a "priority list of health services, ranging from the most important to the least important for the entire Medicaid population". The purpose of the Act was to permit expansion of Medicaid coverage to all Oregonians up to 100 percent of the federal poverty level, and to do so by covering only those services judged to be of sufficient importance or priority. In effect, rather than providing extensive medical coverage for only some of the poor, Oregon chose to provide limited coverage to *every* poor person, as measured by the federal poverty standard.

The most recent version of the Oregon list ranks about 700 medical procedures according to their effectiveness. Depending on how many procedures can be financed from the Medicaid budget, the state will draw a line – paying for every procedure above the line but none below. Initially, Oregon has agreed to underwrite the first 568 procedures – thereby excluding, for example, treatment of common colds and infectious mononucleosis.

There are a number of important features to the Oregon proposal that require attention. First, the Oregon Basic Health Services Act of 1989, if its proponents are to be believed, was meant not as an effort to institutionalize rationing of health care for the poor, but to begin a *process* that will, if successful, establish a unified system of setting health care priorities that will eventually cover the vast majority of Oregon's citizens. There are three separate bills that comprise the approach. The Basic Health Benefits Act expands coverage and access by extending Medicaid eligibility to Oregonians below the federal poverty line. Priorities will be set for preferentially funding services that are judged to be the most effective in contributing to length and quality of life. If funding restrictions occur, procedures with the least potential for benefit will not be offered. The Act will establish statewide managed care through prepaid plans and other mechanisms designed to contain costs while ensuring access and coordination of care. A second Senate bill, the State Health Risk Pool Act, would establish a Medicaid Insurance Pool program for "medical uninsurables", that is, persons who do not qualify for Medicaid and who cannot now qualify for coverage because of preexisting conditions. The Act sets out the ways that state and private insurers will subsidize the pool. Finally, the so-called Health Insurance Partnership Act, if passed, will mandate that four years after the implementation of the Basic Act, employers must provide health benefits that can be purchased through the state insurance pool. The benefits package offered by employers must offer coverage equal to or greater than that provided in the Medicaid benefits package. Any employer who does not provide insurance to all permanent employees and their dependents by a specified date will be taxed at a rate that approximates what would otherwise be the employer's contribution toward insurance.

2. *With Justice For All? The Ethics of Health Care Rationing*

Beyond the general development of health care as a right constrained by common good requirements in the encyclicals, there has also been a recent focused discussion of health care rationing offered by the Catholic Health Association (CHA). The CHA document, *With Justice For All? The Ethics of Health Care Rationing*, provides the most focused analysis to date of health care rationing in light of Catholic principles.

Chapter Three of *With Justice For All?* on "The Public Policy Context" develops eight criteria for assessing the "ethics of rationing". These criteria can be summarized as follows. First, the need for health care rationing must be demonstrable. Second, health rationing must be oriented toward the common good. Third, a basic level of health care must be available to all. Fourth, rationing of health care should apply to all. Fifth, rationing should result from an open and participatory process. Sixth, the health care of disadvantaged persons has an ethical priority. Seventh, rationing must be free of wrongful discrimination. Finally, the social and economic effects of health care rationing must be monitored ([2], pp. x–xi).

Several of these criteria might, at first glance, appear to justify Oregon-like efforts to devise acceptable strategies for Medicaid rationing, while others would seem to disallow such selectively targeted efforts. Although the need for rationing may be "demonstrable", given the rise in health care inflation and the lack of coverage for so many uninsured and underinsured persons, Catholic principles would appear to dictate rationing as a last rather than a first or even intermediate resort. Thus easy or early recourse to rationing as the only practical solution to issues of cost, quality of care, and access may undercut prior efforts to eliminate inefficiencies and waste. Nonetheless, if Oregon's priority-setting is the first step in a more comprehensive process that will lead to systematic scrutiny of the efficacy and cost-effectiveness of medical services, Medicaid "rationing" may work in tandem with practical efforts toward more comprehensive reform.

Moreover, Oregon's plan, again if seen as a first step toward more systematic reform, seems, relative to the status quo, to be oriented toward the common good. As the CHA document notes:

> Clinical medicine must always be oriented to the best interests of every individual patient. However, public policy choices governing the distribution of health care services beyond the level determined to be each person's right must consider the common good. An ethically acceptable rationing scheme must limit expensive medical services, when, necessary, in a way that is fair to all ([2], p. 21).

Again, assuming the best case scenario for the eventual passage of all three bills, Oregon's priority-setting will define for the vast majority of all state residents, beginning with the Medicaid population, what the right to basic care means. Rather than rhetoric, the scope of the entitlement to basic care will be specified. Every poor person, and ultimately all those who, while not poor, are uninsured or uninsurable, will be able to claim a meaningful entitlement. That willingness to set priorities – denying access to low-rated services and procedures while expanding access to cost-effective and efficacious care – appears at least in principle to be oriented to the common good. So too, assuming that the Oregon Medicaid experiment is in fact a first step toward comprehensive reform, the Basic Health Services Act will do much to help establish a basic level of health care for all who are presently marginalized – the poor, the uninsured, the underinsured.

However, Oregon-like proposals are problematic in light of two other criteria set forth in the CHA document. According to the fourth criterion, rationing should apply to all: "Those who construct and implement a health care rationing system are likely better to understand its effects on others if they, too, are required to live within the limitations they construct" ([2], p. 22). And again:

> Only when rationing applies to all can it be the occasion for sharing a common hardship rather than an occasion for deepening the gaps between wealthy and poor, old and young, healthy and sick, and among racial groups. Equity in rationing would suffer if a significant minority of the public obtained their care outside the health care system while acquiescing to limitations on services for those who were economically less secure ([2], p. 23).

This suggestion – that rationing should apply to all – is a criterion obviously at odds with our usual assumptions about two tiers of health care delivery being an inevitable part of health care in the United States. Given the general Catholic discussion of private property and its limits, this criterion is perhaps unsurprising, and indeed, there are two points that might be mustered in its defense. First, the criterion challenges us to reconsider our working assumptions about two tiers of health care. A casual acceptance of two tiers of health care delivery, in respect of the freedom of those better off to obtain more expensive care, may undercut the sense of solidarity that the common good emphasizes. Second, even if the criterion is not implemented, it may encourage us to consider our mutual responsibilities for one another; by so doing, it might serve to broaden the package of basic services that are seen as sufficiently comprehensive, by most citizens, to make "buying out" of basic coverage less appealing. Indeed, in contrast to the U.S. bishops' call for a single-tier of health care, the CHA document discusses a first approximation of satisfying this criterion in a realistic footnote:

> Practically speaking, this criterion could be satisfied if those "buying out" of a rationed health care system constituted a very small percentage of Americans. If too many choose to 'buy out', the rationed system would be disproportionately composed of the economically disadvantaged and would be more likely to decline in scope and quality. A large proportion of Americans "buying out" would also send a strong political signal that the rationed system is insufficiently comprehensive or seriously deficient in other respects ([2], p. 22).

Still, the criterion that rationing should apply to all, even granting the realism expressed in the above footnote, suggests that the Catholic Health Association would deem the present exclusion of certain large groups – especially the Medicare population – as unacceptable, even on prudential grounds. So long as some groups are deemed to be "untouchable" relative to general standards of rationing that apply to other large populations, there emerges a perception, at the very least, of serious inequity that fails to comport to recent Catholic social teaching on property. After all, a large percentage of Medicare coverage draws upon common resources, as with Medicaid. To be sure, there are differences in the initial moral appeals that were made in creating Medicare and Medicaid as legal entitlements: Medicare was, to large extent, seen as a merit-based entitlement due the elderly poor for their years of service, while Medicaid was perceived as an entitlement based on grounds of charity. Nonetheless, though I cannot develop a full argument here, I have concluded elsewhere that Catholic social teaching places moral priority on property in common, based upon use criteria according to need [12]. Consequently, arbitrary exclusion of the Medicare population from the scope of rationing would seem to require something more than prudential judgments to be morally compelling in light of Catholic social principles.

A final criterion might also call Oregon's process into question, the criterion that "the health care of disadvantaged persons has an ethical priority". As with the bishops' resolution, it is difficult to specify the target of this principle; i.e, are we speaking of those generally indigent, or the medically indigent? Although these groups overlap, the principle might well be applied differently, depending upon how that determination is made. One way to consider the issue would be to pose it in Rawlsian terms: does the impact of the Oregon Basic Health Services Act make the worst off better or worse off as a group? On the one hand, the basic Act does not seem to require much from better-off Oregonians. Providers will be reimbursed more for their services. Unless and until taxes rise to expand coverage for Medicaid (which appears quite unlikely), businesses and taxpayers will not pay much more. Moreover, those who are privately covered will continue to enjoy benefits, with tax-subsidized insurance, not available to their poorer Oregonian neighbors. Neither of these results seems in accord with the Catholic emphasis upon the common good and the need for a broader understanding of social justice.

On the other hand, if the Basic Health Benefits Act ultimately functions in tandem with the other two bills, extension of the priority-setting beyond the Medicaid population will do much, as a comprehensive strategy, to improve the lot of all who would gain access, finally, to a meaningful entitlement to basic care. If one reads the future more optimistically, benefits to those presently disadvantaged by gaps in coverage would emerge, and this CHA criterion would be more clearly satisfied.

III. THE RELEVANCE OF CATHOLIC PRINCIPLES TO THE SECULAR DEBATE

In Part I, I discussed the principles of Catholic social thought on medical care as an individual right, but one constrained by considerations of the common good, as well as of social and distributive justice. In Part II, I assessed two recent Catholic documents – one on health care reform, the other on rationing. In this section, I will conclude by posing a series of related questions about how, or whether, a particular theological understanding – here the Catholic tradition on health care – can have resonance with and relevance to the formation and justification of secular policy? What good might a specifically Catholic perspective do in our public discussion of the nature and scope of health care as a right? How might a Catholic perspective further the public debate on such a large and controversial issue as health care rationing?

To revisit an earlier point: we should not assume, at least without argument, that a particular religious voice is irrelevant to policy choices simply because we insist on finding secular and pluralistic warrants for policy formulation. A more full-blooded reading of secular pluralism would celebrate, rather than discourage, the vibrancy of various voices in the public dialogue about difficult issues. James Gustafson reminds us that moral discourse, whether theologically inspired or not, may function in a number of ways. He discusses four "modes" of moral discourse – what he calls ethical discourse, prophetic discourse, narrative discourse, and policy discourse. Each of these, he suggests, is at some level necessary to the moral deliberations of particular communities and society at large but none is, of itself, sufficient [6].

These days, of course, "ethical discourse" may be the mode most familiar to us – the so-called "Georgetown mantra" of principles, the appeals to various rights, the vocabularies of consequentialism and deontology. Ethics is an important language, for it serves to frame our reflections as we justify choices in a pluralistic society where a common narrative cannot be assumed. But ethics, according to Gustafson, tends to be micro-focused, small in scale, working within the status quo, not concerned, much of the time, with the larger cultural or social picture. Ethical discourse may also be rather dry, often trading on technical or legal niceties. Ethics may talk more about the patient's right to refuse treatment than it does about the cultural worship of medical technology that may pose such a quandary in the first place. Ethics may speak exhaustively (and exhaustingly) about the autonomy of individuals in making their own decisions even as it gives short shrift to the discussion of the individual's responsibilities to the larger society.

Prophetic discourse, by contrast, is often passionate in its sweeping indictment of larger cultural trends and social sins. Such discourse makes up in vision what it lacks in precision. It forces us to notice the forest through the trees, to see those large-scale background features that the ethical mode in the foreground often tends to underplay.

The discourse of narrative is the language of story, the story that shapes a community – in the Christian tradition, the centrality of the good news, the complexity of the parables, the attitudes and ethos shaped by the story the community tells, the faithfulness of the community to its own shaping narrative. Narrative is not the language of argument, of precise moral reasoning, of premises and conclusions. Narrative is more full-blooded, less skeletal. The bare bones of argument so dear to secular philosophers are covered with the flesh and blood of tradition. Narrative, before all else, is about inspiration, about the formation of character. We are shaped by the stories that we tell.

Finally, there is policy discourse. Policy discourse, rather than focusing on ideal theories or grand conceptions, usually functions within the constraints of history and culture. It works with the values already embedded in the choices that we have made. It seldom, if ever, is prophetic; its horizon is limited. It generally asks not, "What is the good or the right choice?" but, within a range of alternatives, "What is the reasonably good and feasible choice?"

It is useful, in light of Gustafson's distinctions, to consider in which mode or modes Catholic social teaching belongs in debates about health care reform and rationing. With all the appropriate caveats in place concerning the differences between the formation and the justification of public policy in a secular pluralistic society, Roman Catholic discussions of health care might contribute to policy formation in a number of different, perhaps complementary, ways. As ethical discourse, Catholic thought might call us to achieve a better balance between the language of "rights" and "obligations". Individuals have rights, but the language of the common good provides a useful counter to the usual stridency of rights language. As prophetic discourse, the Catholic understanding of the limits on private property might call into question the unexamined assumptions of a secular society about the seemingly sacrosanct status of individual acquisition and private property. Moreover, the CHA criterion that rationing should apply to all – however impracticable it may be – might force us to reconsider our status quo assumptions about separate and unequal forms of care.

So too, the Catholic voice, by emphasizing certain themes in the Christian story – especially the universalizing tendency of Christian love – might invite broader reflections about what we owe one another and how we are responsible to one another for the meeting of our basic needs. In addition, the richer narrative voice would surely reinforce distinctive emphases in Catholic institutions, perhaps about the sorts of services that they will or will not ration. Finally, the Catholic policy voice, proclaimed by the bishops in their resolution and by the CHA in its statement on rationing, will challenge Catholics, and perhaps others, to consider their own best moral instincts and values, especially those already embedded in earlier policy choices about expanding access and improving care, as exemplified by the passage of Medicare and Medicaid.

None of these modes of moral discourse, as I have suggested, is likely to provide, in itself, a specifically "Roman Catholic" warrant for secular social policy or public choices on health care reform. Indeed, in the two documents that I analyzed in Part II, the mixture of modes is quite striking, even as it leaves one uncertain about how to interpret the authority of particular conclusions. However, because Roman Catholic social ethics continues to invoke natural law categories of reflection, one would not expect Catholic policy discourse to be unique (although on some issues it will retain a distinctive cast). After all, the strength of natural law, in principle, is its availability as a source of natural moral insight to persons of good will. Yet it is worth remarking, as I close, that the persuasiveness of natural law in Catholic arguments has been quite variable, even to those still somewhat sympathetic to that methodology. As Bryan Hehir observes, recent Catholic policy statements argued in natural law terms have been received more favorably on general matters of social ethics than on specific issues in bioethics and sexual ethics. For example, in their pastoral letters in the 1980s on the economy and war and peace, the U.S. Catholic bishops invoked a recognizable version of natural law. In Hehir's judgment:

> Even though such a position is not well known or widely used in American academic or social policy debate, the positions of the bishops catalyzed a broad public discussion and found support beyond the boundaries of the Catholic community. [Moreover] [t]he bishops' philosophical perspective often found more support than the specific policy conclusions they drew from it ([7], p. 357).

By contrast, on particular biomedical and sexual issues, "it has been very difficult to get a hearing for either the philosophical foundation or the conclusions espoused by the bishops" ([7], p. 357).

The reasons for that difference in reception are complex, but they involve, most crucially, the way that general principles and particular policy conclusions are distinguished or conflated in social ethics and bioethics respectively. The counsels of prudence usually feature more prominently in social teaching than in the strict moral conclusions traditionally reached on sexual and bioethical issues, primarily, as Hehir notes, because the latter involve judgments about "intrinsically evil acts" ([7], p. 358). To be sure, the recent stance taken by the U.S. bishops against support for any package of basic health care benefits that includes abortion exemplifies a point of intersection between two ordinarily distinct approaches. Indeed, the policy arguments required to sustain that position will offer an instructive challenge for Catholics who wish to speak appropriately in the policy mode. While one would clearly expect religious values to be invoked as focal elements of prophetic or narrative discourse, policy arguments, on abortion and other issues, require nonparochial, i.e., public, reasons. This, indeed, remains the central drama of being a public church: to witness by persuasion as well as by example, to speak to similarities as well as differences, to discover and celebrate those commonalities of experience and reflection that allow religious values, albeit indirectly, to work their leaven upon the world.

Institute of Religion/Center for Ethics
Texas Medical Center

NOTE

1. Significant sections of this essay closely parallel my analysis of recent Catholic documents on health care in 'The Common Good in a Secular Society', *The Journal of Medicine and Philosophy* 18 (1993), 569–587.

BIBLIOGRAPHY

1. Ashley, B.M. and O' Rourke, K.D., 1978, *Health Care Ethics*, Catholic Hospital Association, St. Louis.
2. Catholic Hospital Association (CHA), 1991, *With Justice For All? The Ethics of Health Care Rationing*, St. Louis, Mo.
3. Christiansen, D., 1991, 'The Great Divide', *Linacre Quarterly* 58 (May), 40–50.
4. Engelhardt, H.T., 1986, *The Foundations of Bioethics*, Oxford University Press, New York.
5. Gremillion, J. (ed.), 1976, *The Gospel of Peace and Justice: Catholic Social Teaching Since Pope John*, Orbis Books, Maryknoll, New York.
6. Gustafson, J., 1990, 'Moral Discourse About Medicine: A Variety of Forms', *The Journal of Medicine and Philosophy* 15 (April), 125–142.
7. Hehir, J.B., 1992, 'Policy Arguments in a Public Church: Catholic Social Ethics and Bioethics', *The Journal of Medicine and Philosophy* 17 (June), 347–364.
8. Hollenbach, D., 1979, *Claims in Conflict: Retrieving and Renewing the Catholic Human Rights Tradition*, Paulist Press, New York.
9. John Paul II, 1981, *On Human Work*, United States Catholic Conference Office of Publishing Services, Washington, D.C.

10. Keane, P., 1993, *Health Care Reform: A Catholic View*, Paulist Press, New York.
11. Langan, J., 1986, 'Common Good', in James Childress and John Macquarrie (eds.), *The Westminster Dictionary of Christian Ethics*, Westminster Press, Philadelphia, p. 102.
12. Lustig, B.A., 1990, 'Property and Justice in the Modern Encyclical Literature', *Harvard Theological Review* 83, 415–446.
13. Outka, G., 1974, 'Social Justice and Equal Access to Health Care', *The Journal of Religious Ethics* 2 (Spring), 11–32.
14. Treacy, G.C. (ed.), 1939, *Five Great Encyclicals*, Paulist Press, New York.
15. United States Bishops, 1993, 'Resolution on Health Care Reform', *Origins* 23, 98–102.

DAVID C. THOMASMA

The Post-Modern Challenge to Religious Sources of Moral Thinking

A rupture in the orderly development of moral thought has occurred largely in this century that makes a qualitative difference in the way we deal with moral analysis. This rupture, which is often called the post-modern or post-industrial reaction to the fundamental ideas of the Enlightenment, creates difficulties for reaching any kind of moral consensus, be it secular or religious. Post-modernism is not just a recognition of pluralism, or even a kind of ecumenical openness to a wide range of ideas; it is rather a direct assault on the very possibility of reaching a consensus on ethical issues. The sources of the assault on this possibility are many and complex. Recounting some of them forms the first part of this essay.

In fact, as MacIntyre's 1988 Gifford Lectures demonstrate, there are at least three modern forms of moral enquiry in which moral debates occur and between which moral conflicts arise. Not only are the differences among the structures and rules of different moral enquiries vast and irreconcilable, but also many of the debates among adherents within such systems, e.g., Aristotelian ethics, deontologism, Nietzsche's enquiry, the moral analysis found in an Encyclical by a Pope, and so on. MacIntyre observes:

> . . . debate between fundamentally opposed standpoints does occur; but it is inevitably inconclusive. Each warring position characteristically appears irrefutable to its own adherents; indeed in its own terms and by its own standards of argument it *is* in practice irrefutable. But each warring position equally seems to its opponents to be insufficiently warranted by rational argument. It is ironic that the wholly secular humanistic disciplines of the late twentieth century should thus reproduce that very same condition which led their nineteenth-century secularizing predecessors to dismiss the claim of theology to be worthy of the status of an academic discipline ([47], p. 7).

MacIntyre's description of irreconcilable moral enquiries, while accurate and interesting, tends to overemphasize differences rather than the unifying features of human character and the moral life. To be sure, there are disagreements in principle, in method, in the weight given to history and tradition,

51

E. E. Shelp (ed.), Secular Bioethics in Theological Perspective, 51–74.
© 1996 *Kluwer Academic Publishers. Printed in the Netherlands.*

and in what counts as a good argument. But within those differences, people may behave similarly with respect to compassion for others, keeping promises, being courageous, and acting prudently. The virtues, therefore, as elements of character, may transcend differences in modes of moral enquiry because they are ultimately related to what it is to be fully human. The effort to seek a morality grounded in our common humanity should not be abandoned, notwithstanding the current incommensurability of notions of the good [56]. In any case, I believe that in the limited case of medicine, it is somewhat easier than in general discussions of ethics to find agreement on the good, on ends, and on purposes.

That being said, there is no call for a rosy assumption that even within a tradition of moral enquiry there will be agreement, as Engelhardt has shown [23]. First of all, international and intercultural differences are very hard to "factor in" to any religious tradition of moral enquiry [46], [67], even those that are representative of a universal or world-wide church. Second, and even more important, is the recognition that the same post-modern forces have had an impact on religious sources of moral teaching too, especially the teaching authority of the church. Note how often synods or councils of various Christian church bodies have difficulties reaching a consensus about statements on nuclear warfare, population control, abortion, empowerment of the poor in underdeveloped countries, pre-marital sex, the ordination of women, recognition of Gay and Lesbian unions, and more to the point of medical ethics, the status of the embryo in research and transfer, reproductive rights, transplantation, withdrawal of fluids and nutrition, euthanasia, the role of religious health care institutions, and even cooperation between rival religious healthcare systems!

In constructing this argument about our disjunctive age, I will be concentrating on the Roman Catholic tradition. Furthermore, I will try to illustrate my points by using bioethics issues in a sphere of moral enquiry that argues more often from tradition and authority than other spheres of moral enquiry. In principle this should mean that there would be more *a priori* cross-cultural and transhistorical agreement. But as we shall see that is not possible even within a single sphere of religious enquiry because of the realities of the post-modern world. My thesis is that post-modern, secular reality, requires post-modern methodology in religious or theological thinking.

Rather than reject post-modern thinking, I try to incorporate its secular reality into a critique of traditional methods of reasoning in Catholic moral thought. First I explore the markers of the post-modern era, second I look at religious sources of moral thought in the Catholic theological tradition and the need for an historical hermeneutics, third I examine the conceptual embarrassment caused by "lifting" ideas from the past and applying them to the present without hermeneutics, fourth, I examine the role of lived experience of believers in grappling with bioethical issues, and then I draw some conclusions about Catholic moral reasoning today.

1. Markers of Post-Modern Era

What is it about our age that makes it "post-modern"? There are certain markers of our age that set us off from prior periods. As progress increases so do the problems of alienation that heightened and rapid progress brings in its wake. All of the following, non-exhaustive features of our age, contribute to social and historical disruptions.

1.1. *Freud and Psychotherapy*

Freud's postulation of the subconscious and unconscious contributes a sense of loss of moral responsibility in our age. In August, 1994, when Robert Aller, the father of a schizophrenic, confronted the vast system of research on the mentally disabled as exploitative and lacking even a modicum of adherence to the Helsinki Accords on medical research, his son's psychiatrist went on the "Larry King Live" television program and said that the father was just angry that his son had schizophrenia. The father's moral objection was reduced to a psychological defect. That way, no one has to take responsibility, except, of course, the angry father.

Furthermore, when what we normally would consider to be a free action legitimately can be analyzed into components that drive the subconscious, then that action is not as free as once assumed. If an eighth-grader forces a girl to have sex because repeatedly he saw something like that on TV, how responsible is he? And if that girl later mistrusts men so much that she goes through three divorces before she confronts her own agonies from the past, is she responsible? How strict can religious moral teaching be in this kind of environment? Motivations are surely shaped by the social and cultural environment. Speaking about moral issues in such an environment requires a balance between acknowledgement of subconscious forces and standards of behavior that, in other ages were taken for granted, but now might be almost supererogatory, given the pressures on individuals and communities today. Thus, the moral voice is muted.

1.2. *Interlocking Technology*

Not only are we a technological age. We are also in an age of far-reaching, interlocking technologies. Oil, plastics, textiles, computers, automobiles, are all interrelated and interlocking. To pull the rug out from one technology affects all the others. Coincidentally, it also affects our sense of personal and individual autonomy, as well as social cohesion [70].

Some examples: Rapid transit systems between our cities (which we need) would decimate the automobile companies and workers, throw entire states into receivership, and destroy workers and companies in the oil industry, gas stations, roads and bridges, traffic lights, and so on. The Valdez tanker that ran aground in Alaska, reportedly because the captain was drunk, created major interlocking problems [75]. Before the enormity of that spill was recognized, gas prices throughout the country rose at the pump by 10 cents, and Alaskan salmon became endangered along with the rest of the fishing industry. Even the sea otters needed to be rescued from freezing, their coats having been exposed to oil and to the detergents to remove it. Still another example of interconnectedness: the disinfectant used in American hospitals cannot be employed in China because the Chinese sewer systems are not able to handle it [36].

Because of this interlocking character of our technology, we have lost some control of its direction and implementation in a significant way not found in the past [68]. Ideas cannot be mandated and implemented without careful human impact reflection, consulting not only the affected persons in our own culture, but also those in other cultures and traditions. Dialogue about moral issues must occur before the fact, not after a new technology is introduced.

1.3. *Economic Codependency*

Not only are technologies interlocked. All persons are interlocked in an economic co-dependency that, as yet, has no international political expression. Recall several years ago when two red grapes were found to be injected with cyanide, how the U.S. shut down imports of all fruit from Chile. The communists were blamed. The economy of Chile suffered an immediate blow of $100,000,000 loss. Many people were put out of work. Chileans blamed U.S. government interests in trying to bring down then-president Pinochet. This happened at a time when all of South America suffered under a huge, unfathomable world-wide debt. The debt led to inflation rates of 1,000 percent and crackdowns on the economy that in many countries led to riots [4]. American retailers were hurt, also, by such a shutdown, as were school cafeterias and delis.

The debate about NAFTA (the North American Free Trade Agreement) early in President Clinton's administration (1993) demonstrated the realities of our interconnectedness as well. Agreements like these are just the beginning of a political cooperation that will follow the economic one. Thus, even though we may profoundly disagree with China about human rights, America recently put those objections aside for billions of dollars of trade that will inevitably alter their political system (and our own). Thus, the economic interconnectedness ironically leads to a silencing of the moral voice in favor of long-term benefits that may bring about changes later. The same sort of negotiating occurred at the 1994 UN-sponsored population and development conference in Cairo. There, the Moslem countries and the Vatican opposed wording in a draft about reproductive rights that seemed to approve of abortion. Political compromises were reached eventually in favor of the important development agenda.

In the post-modern era, such skills are not optional. They must be part of the ethical methodology.

1.4. *The Power of the State*

The state is the major reimbursement mechanism for caring for the elderly and poor, through Medicare and Medicaid. In each story of the collapse of yet another hospital, the blame is put squarely on the state. Some states, like Illinois each year in April, run out of Medicaid funding. In one year this caused a number of inner city hospitals, almost 75 percent reliant on Medicaid for reimbursement of the care they gave to the poor, to close. By contrast, the Catholic Health Association has long endorsed the doctrine of the preferential option for the poor. This means that, should two persons arrive at the emergency room, one poor without reimbursement possibility, and the other rich, the facility should choose to care for the poor person first. How could such a remarkably beautiful moral commitment be made in a culture created in the image and likeness of competition and inadequate state reimbursement for care?

Efforts to require the state to practice justice are only just beginning in our calls for health care reform. In fact, these efforts have been pre-empted by both economic and political considerations that make a mockery of the ethical underpinnings of Clinton's original plan. The voice of religious ethics should be heard loudly and clearly in this effort. Yet the Catholic Church, supporting universal coverage, backed off supporting current plans because they provide for abortion at public expense. Once again the moral method is meshed with political agendas.

1.5. *The Power of the Media*

The media reduces all issues to tiny "spots". As Sidney Callahan observed, "TV favors 50-second ethics, with clear, punchy, controversial opinions. The medium does not reward introspective self-doubts or displays of complex moral thinking" ([13], p. 20). And the media can become involved in the issue, influencing the outcome itself. The media can shape conclusions families and society reach through its intensive questioning and even accusative power. Or the media can supply public access to health care for some who would not normally obtain it. Consider the following two examples below.

In Chicago in April, 1989, Mr. Rudy Linares disconnected the respirator from his son, who suffered from a permanent vegetative state as a result of swallowing a balloon, and held him in his arms while the son died. He did so by holding off health professionals with a gun. The media made him a *cause célèbre*. Media and telephone surveys showed that the populace favored 13 to one not indicting Mr. Linares for murder. Eventually the grand jury failed to indict him because both he and his son and the rest of his family were held victim by a rampaging medical technology. Later a law was drafted by a State's Attorney commission formed to study the problem of withdrawal of care. This was the "Illinois Surrogate Decisions Act", one that provided for withdrawing medical technology from persons without capacity to make decisions for themselves. The media had an important role to play in this favorable outcome [29].

Similarly in Chicago, at Loyola, the Lakeberg twins were born at the end of June 1993, conjoined at the sternum. Prior to their birth, the parents, who were a poor family from Indiana, notified the media. Throughout the care of the twins, the parents went to the media constantly, thus essentially making it impossible for a reasoned consideration of withdrawal of care. In fact, the father directly pleaded with the media after Children's Hospital of Philadelphia agreed to separate the twins (widely regarded as a medically futile and expensive effort) for a "made for TV movie" about all this. He had hoped to get money from their plight. During call-in shows, parents of children who had a much greater chance to survive, say from a liver transplant (that has a 50 percent five-year survival rate), cried, complaining that they did not know how to capture media interest for their children so that they, too, could be saved [73]. The media's role in this case was less favorable, and partially may have driven the desire to save the children.

1.6. *The Modern State*

Nazi Germany is widely regarded as the first example of the power of the modern state to control technology, information, economics, and thereby its citizens. A more recent example of this power of the modern state appeared during the Rushdie affair. Salman Rushdie wrote *The Satanic Verses*, highly critical of the Prophet Mohammed, and sexually offensive to Moslem leaders. He was condemned to death by Iran and the Ayatollah. This led to international withdrawals of embassies by Western European countries. Internal debates about the freedom of the press and of expression occurred in England and the United States especially. Workers at comfortable mall-based book stores were threatened by calls and bomb scares. The book was "kept in the back". Full-page ads were taken out by Waldenbooks defending their decisions about the book. Similar ads were taken out by Moslems, defending their religion and their religious sensibilities. A similar occurrence revolved around Taslima Nasrin of Bangladesh, a physician nonbeliever critical of Moslem views of women and marriage. She is now hidden in Sweden [77].

Much of the reflection on the Rushdie and Nasrin affairs betrayed a note of Western cultural superiority. We admonished fundamentalist Moslems that they must enter the golden era of responsible international citizenship. Amnesty International took up their causes. Protests were lodged by Western Governments with Iran and Bangladesh. Some countries threatened to cut off economic assistance. Yet not that long ago, the power of Christianity was allied to the state for almost 1,000 years, during which there were many persons who were tortured and killed, put on racks and burned at the stake, for being different or refusing to follow the party line. We escaped the "medieval and violent darkness" of Christianity, in the words of one commentator, "by depriving Christian religious authorities of political and legal power over the community" [22].

Even though we must be cautious of the tyranny of the modern, secular state, its benefits in terms of individual freedom of expression are obvious. Hence, we rather happily no longer share the basic, social common denominator of medieval society (and I daresay, of fundamentalist movements anywhere), a common faith that is reinforced in all social structures. The Christian church only reluctantly has come to grips with the fact that it cannot control those structures as it has in the past. It must, instead, seek to persuade by the goodness of its deeds, so that its words carry a moral authority missing in other realms of society. This alters the moral voice from one of command to one of persuasion.

1.7. *The Nuclear Background*

The destructive force of nuclear warfare is another essential background that leads to disjunction in our society. The incredible dependency of foreign and even domestic policy on the absurdity of nuclear warfare is a kind of cultural insanity that eats away at all our rational efforts to be moral. Witness the economic recession in California that resulted from the collapse of the Russian Empire, and the loss of enormous revenue in the military-industrial complex. In health care, constant reference to the billions of dollars that are "wasted" in defense impede plans to offer health care to the 38 million citizens currently under or uninsured, who cannot get access to health care. Rational efforts shipwreck on the essential injustice of using money for life-destructive goals rather than life-enhancing goals.

Ironically, the nuclear race bequeathed something even more profound than the anxiety that the world will end with a bang. Sometimes postmodernism requires us to turn tables on sanity itself. As Saul Bellow says through Henderson, the Rain King: "To be sane in an age of madness is also madness" [11]. Thomas Merton also shared this insight:

> I am beginning to realize that "sanity" is no longer a value or end in itself. The "sanity" of modern man is about as useful to him as the huge bulk and muscles of the dinosaur. If he were . . . a little more aware of his absurdities and contradictions, perhaps there might be a possibility of his survival. But if he is sane, too sane . . . perhaps we must say that in a society like ours the worst insanity is to be totally without anxiety, totally "sane" ([50], p. 49).

If moral analysis lacks this almost bizzare, anti-rationalist tinge, then it seems to many to be "out of touch". It seems almost too serenely confident of its own truth.

1.8. *Evolution*

The theory of evolution most certainly contributes to both our historical under-standing and the contextualization of modern thought, as well as to our awareness of the disjunctive properties of current cultural experience. For the most part, we assume that culture itself has evolved, and that our current state of affairs is "better" than that enjoyed by people in the past.

But the hermeneutics of caution must also be applied to any notions of traditional thinking. The problematic of past sources is not their problem, but ours. Disrespect for tradition in moral analysis is a mark of our times that leads to either a loss of continuity when one accepts it, or an idealiza-tion of the past, if one reacts against it.

1.9. *Summary*

These and other features of our civilization lead to inevitable conclusions about the ethical challenges of our age:

1)   Despite our optimism about human progress in the past, our depen-dency on technology, the state, and the media diminishes human freedom in a significant way. The result must be incredible caution about human progress.

2)   This caution must be exercised especially with regard to the goals of science. Science itself cannot suggest the proper goals of a human civiliza-tion. The issue is joined at the difference between what can be done and what should be done. Joseph Fletcher asks about genetic engineering if there is a socially responsible science today. He ruminates:

> Whenever I read news stories about "activists" who demonstrate against nuclear weapons or nuclear power installations, or express outrage about toxic waste hazards, or allege the dangers to society and human life of research and development on frontiers such as biotechnology or ABC warfare (atomic, bacterial, and chemical), I cannot just dismiss it all as emotional imbalance. Instead a web of wonder starts to spin itself ([24], p. 565).

Fletcher concludes that there are moral limits to knowledge, that values ought to outweigh truth, when that truth can be put to destructive use.

Gaining control over those goals of science and technology is the most important political and moral task of the 21st century. As E. G. Mesthene observes:

> New technologies by their very nature must challenge existing social values and institutions. The opening up of new options for human action must call the old ones into question. As man gains control of the process of change, he is forced to decisions on the ends to which he will direct his own future ([47], p. 135).

3)   As our power over nature increases, we increase our power over

ourselves as well. To seek to control the interlocking processes just adum-
brated, is to seek to control our own biology, ecosphere, society, and culture.
Leon Kass starts his book on biology and human affairs with this observa-
tion:

> We belong to nature naturally, we place ourselves outside of nature to
> study it scientifically, in part so that we may be able to alter and control
> it technologically. Yet because we belong to the nature we study and seek
> to control, our power over nature eventually means power also over our-
> selves. We are not only agents but also, and increasingly, *patients* of our
> scientific project for the mastery of nature (author's emphasis) ([39],
> p. xi).

4) We must acknowledge the essential plurality and relativism of our age
with respect to other times in the past. We inherit different methods, different experience, different proclivities, and we address different audiences.
Fortunately faith itself can be enriched by this diversity. For one thing, we
can become much more sophisticated about levels of moral concern, e.g.,
how helping a couple become fertile using medical technology is at a much
higher level of moral action than aborting a fetus at 22 weeks because the
pregnancy is unwanted.

Another result of the recognition of plurality and relativism is a better
articulation of the context in which certain moral truths apply compared to
another context. Alasdair McIntyre argues quite effectively that without
unanimity in principle, ethics is condemned in our time to bickering between
assertions matched by counterassertions. He does think, however, that communities of concern can create within themselves certain moral principles
that would apply to that concern [43]. An example might be the axioms that
Edmund D. Pellegrino and I articulated from the goal of medicine to heal.
One cannot heal if one destroys the autonomy of the patient in the process
[55], does not right the imbalance of dependency and power between the doctor
and patient, and does not treat each person as a class instance of the human
race ([54], c. 7).

5) The language we use can be confusing. Even the term "moral theology"
might mean one thing to Cardinal Ratzinger, another to a young woman
theologian working in a small Catholic college in Pennsylvania, and still
another to a fellow working at the Hastings Center.

With these reflections about the post-modern age in hand, I now turn to
the "tradition" itself to see if there is a realistic basis upon which it can assist
persons in such a world with some degree of moral authority.

2. Religious Sources of Moral Analysis

2.1. *The Period of Historical Naiveté*

The usual appeal to scripture and tradition in dogmatic and moral thought causes us problems. We are much more aware of the problems since the rise of historical consciousness in the 19th century. Prior to that time, statements of the past were applied in the present without an explicit hermeneutic. We might call this period, the period of "historical naiveté". Sufficient continuity with the past permitted scholars like St. Thomas Aquinas to grapple with the sayings of scripture and the fathers of the church using the scholastic method of reconciling conflicting statements. Apparently the origin of this method lies in early medical texts of the Arabic period, wherein opinions of different physicians regarding a symptom or disease process would be listed, and the effort made to reconcile these. The method was refined in the Middle Ages, finding its way into public debates (the *Quodlibets*) and the structure of the articles in the *Summa*, as well as into the courtesy extended to other thinkers, namely, that one would never deny the major premise, but would distinguish the minor [57]. Sometimes this propensity serves scholars well who wish to argue against critiques of traditional moral thought, say in a debate about whether one must love one's neighbor equally as well as oneself (the Impartiality Thesis) [64].

Nonetheless the scholastic method is an ahistorical one. It employs, for the most part, a logical hermeneutic. It serves moral thinking well, but lacks any real effort to determine the context of the prior statements, the audience, the concerns of the time, the specific challenge that prompted the original response. Most often there was no way that scholars until the last century could interpret the historical sources anyway. In prior times, historical analysis was not yet properly conceived. Indeed, the very idea of history was quite different than our own. The story itself, and its moral point, was more important than accuracy.

It would not be right to claim that no historical awareness existed in the middle ages. As Nicholas Lobkowicz, in his *Theory and Practice* argues, Christianity, not antiquity, laid the groundwork for the modern notion of progress. Antiquity was enamored of cyclic seasons. Christianity, based on the Jewish experience of salvation, introduced the idea of an historical event that never before existed and never will again ([41], p. 98). Thus, even an ahistorical thinker like Aquinas was aware of the development of theoretical thought and of the realm of *operabilia* useful for human society [*Summa Theologiae*, II-II, 1, 7 obj. 2; II-II, 97, 1]. But because of the rediscovery of Aristotle and other major thinkers of the past, the 13th century was less able to grapple with the development of thought than the 11th and 12th had been. It concentrated, by contrast, on obtaining accurate translations of these sources. Earlier, however, Hugh of St. Victor saw historical development as a human response to the wretchedness of evil and sin:

Therefore he [humanity] began to struggle for liberation in order to overcome evil and become good. This is the origin of the quest for wisdom which one calls philosophy. It is sought in order that ignorance becomes illuminated by a knowledge of truth, evil concupiscence extinguished by the love of virtue, and bodily infirmity reduced by the discovery of useful things ([10], p. 151).

Robert Kilwardby, the Dominican scientist at Oxford University one hundred years later held a similar view, that human reason was given in order to overcome obstacles so that we could clothe and feed ourselves, for nature had not provided the easy means that it had to animals and plants in this regard ([63], especially p. 18). Still the notion of the inherent development of both doctrine and morals was lacking.

For our considerations this means that all thinking before the 19th century contains a fundamental flaw, precisely because it does lack a proper sense of historical context for its own thinking. This does not mean that such thinking cannot reveal the truth. Nor does it mean that people in prior times were less able to resolve moral conflicts. The opposite may in fact be true. Thus, while we must be diligent about interpreting the past, we should not claim any moral superiority to it.

Rather the lack of historical consciousness in earlier thinking means that we must exercise what feminists call the "hermeneutic of suspicion" on all prior theologies [38]. Ditto for the teachings of the church that arise in this time. A good example of using such caution-hermeneutic is that employed by Eduard Schillebeeckx in his analysis of the meaning of the real presence in the eucharist in, *Christ: The Sacrament of the Encounter with God*. The real presence is affirmed, but the Aristotelian interpretation of the reality of that presence is not ([61], pp. 86–119). Note, too, how Karl Rahner interprets obedience in the church. There is an effort to affirm its importance. After this is accomplished, the "meaning" of obedience is discussed. This meaning, of responsiveness to the spirit, is not the kind of authoritarian meaning customarily associated with the notion of obedience in the church in the past [59], [60]. It has been reinterpreted for our times. This leads me to the second era.

2.2. *The Period of Historical Consciousness*

The second era might be called the era of "historical consciousness", one in which one must detail an explicit moral hermeneutic about any past thinker or teaching, as Hegel and Croce demonstrated. In this era thinkers were aware of the hermeneutical problems of applying the past to the present. The problems arise from the "embarrassment" of prior statements made in an historical context, a context with which we must grapple, out of context, in order to properly understand such statements. The problems are caused by our desire, no, commitment to take the past seriously, and to respect fellow pilgrims in faith. These fellow-pilgrims in faith in the past were challenged by their times as we are by ours. They are part of an historical faith-community.

It is this commitment that connects Christian morality today to wandering Arameans in the desert 3,000 years ago, to prophetic and wisdom thinkers of the Old Testament, to rabbis and apocalyptic visionaries, to fishermen and apostles, Greeks and Romans, Irish monks, and indeed, to international thinkers and doers for centuries. Christian moral thinking is cross-cultural and trans-historical. In this respect it is truly "catholic", i.e., universal.

The point about commitment is important. In the early church, when dogma was contained in a single credal sentence about the death and resurrection of Jesus, moral thinking still occurred. But it was based, as was the *kerygma* itself, on the *experience* of the church. I underline that, and will return to it later. The early debates about the humanity of Jesus could not be solved by appeal to a creed. Instead, the experience of those who had been redeemed compelled them to argue that Jesus was fully human, since every aspect of human life was redeemed. St. Paul's argument against consorting with prostitutes was derived from the experience of "belonging to Christ," that the body was redeemed now and belonged in a mysterious way to the body of Christ.

Recent Protestant moral thinkers move more directly than Catholics from scripture to the present, using one or another current philosophy to reinterpret their experience. Paul Tillich's moral thought or H. Richard Niebuhr's or for that matter Harvey Cox's spring immediately to mind. Tillich relied on existentialism [74], Niebuhr on sociology [52], and Cox in his earlier and most famous work, on a secular hermeneutic [19]. The way Protestants think about morality is more often parallel to the thinking in scripture itself, from experience filtered through a particular philosophy (St. Paul used Stoic categories of virtues and vices, for example), to a confession of faith.

However Catholic and Protestant thought, indeed, philosophical thought in the past 100 years, has been dominated by "the search for method". Our bookshelves are lined by volumes unearthing truth's relation to method. Think of Karl Löwith's efforts to trace meaning in history [42], Wilhelm Dilthey's explorations of the relation of history and society [21], Mandelbaum's answer to historical relativism [45], Heidegger's lifetime efforts to enmesh being in time, and to understand human relationships to both being and to science [31], Gadamer's hermeneutics [27], and Robert Funk's combination of linguistic analysis, new criticism, and hermeneutic into a "rigorous new theology" [26]. These are just a few major theoreticians trying to find some valid viewpoint with which to address the past. This enumeration is a clue to the dilemma we face as we move into the 21st century. Because of the post-modern challenge, it is no longer sufficient to assume a continuity with the past by which we can "reinterpret" it in the present. In addition to an accurate representation of past thinking (if that is possible), and an explicit discussion of the hermeneutic by which we interpreted the past thinker, something additional is now required.

The context of the original statement must be interpreted, and we must defend the very applicability of the insight carried over into the present. The defense is required because of the disjunctive nature of our own experience. This point leads me to the third era.

2.3. *The Era of Historical Disjunction*

If the first era was one of historical naiveté, and the second one of explicit hermeneutic, what is the nature of the third? It is this one, the era of "historical disjunction", that has occupied us as the post-modern crisis in moral analysis.

Karl Rahner wondered in his essay on free speech in the church why people pay so little attention to moral statements by believers:

> We have, too, an unpleasant sense, whenever we hear the sound of our own voice, that it is not particularly surprising that nobody listens to us. Doesn't a great deal of what we say sound strange in our own ears – outmoded, utterly out of date? It is still lodged in our heads, as a hangover from the days when we learned our catechism, but it no longer comes from our hearts: so it is hardly to be wondered at that it no longer gets into the heads of other people ([58], p. 55).

The disjunction between past and present requires a new hermeneutic, a new method of dealing with current experience. We cannot simply state "truths" from the past, especially in an authoritative manner, if those truths are not experienced in the present or open to the future. The "something new" that is required by our disjunctive age is a kind of attention to "raw experience" in the present from which the past can be interpreted and the future examined. In this regard, Thomas Merton asks: "What is a Church after all but a community in which truth is shared, not a monopoly that dispenses it from the top down"? ([49], p. 152)

Most scripture scholars, by way of another example, recognize that one must describe the context of any ethical utterance in the Bible. Yet this hermeneutic is widely ignored by preachers and theologians. R. Joseph Hoffmann postulates that the reason is "because recontextualizing the isolated aphorisms of Jesus and Paul has the undesired effect of marking out the distance between what contemporary believers think Jesus said – as interpreted by the medium of church and Sunday School – and how his sayings were understood by people of his own time." ([35], p. 61) Efforts must be made to reconstruct the communities' beliefs, the groups that remembered and set down the sayings. The disjunction between those communities, communities of elective poverty and pacifist tendencies, and our own faith communities is enormous. It creates a "halting" effect to all analysis and commitments, as if the moral voice has developed a profound stutter.

Thus, the historical disjunction is one of the community's current experience compared to the past. Debates about the meaning of past moral teaching rarely descend to this level of recognition. Charles Curran, to take a recent example, was not just reinterpreting past moral teaching in the church; he was also reflecting on a new experience of the community in the present, one that is so powerful that it cannot be denied, despite all the pain and suffering that questioning past teaching can cause. In the end, the community that is the church accepts or rejects moral teaching commensurate with its experience.

Let me look a little more deeply at the embarrassment of past statements.

3. THE EMBARRASSMENT ANALYZED

How to honor one's faith doing bioethics in a post-modern society? Should we, as Pope John Paul II suggested, "guard" that faith and keep it pure in such an age? In his closing remarks to American bishops at a meeting to reconcile differences with the Vatican, the Pope said:

> We are guardians of something given . . . something which is not the result of reflection, however competent, on cultural and social questions of the day, and is not merely the best path among many, but the one and *only* path to salvation (author's emphasis) [34].

This model of faith seems too Platonic. It assumes that we have an unbroken reach into the past. Faith is seen as a treasure, unchanged by place and circumstance, beyond human reflection and human orchestration, beyond the tatters of time's clothing. Is this a realistic and accurate representation of faith? The question itself is not new. Cardinal Newman raised it in the last century.

Should theological bioethics instead immerse itself in our culture and leaven it? At the opening of the same meeting, Archbishop John May of St. Louis noted, for example, that the authoritarian model of the teaching authority of the church to Catholics in the United States has as much appeal as "the divine right of kings" [6]. This view assumes that faith is interlarded in human life. Just as human life and culture progress, so too does moral teaching. The divine right of kings is a political teaching that has been surpassed by democracy. Authoritarian models of teaching authority in the church have been supplanted by post-Vatican II dialogical ones, ones incidentally, that work well in American society. Archbishop John Roach told the Vatican officials and the Pope not to view as a weakness the American Bishops' decision to follow an "American tradition of dialogue and compromise" [32]. Dialogue and compromise is experienced by the faith-community in the United States at least, as an advance on the model of teaching authority developed by the church in a different social milieu.

There is a trust in the faith-experience of those closest to the problems. For example, if Catholic health care institutions, using the best models of medical decision-making of our time, ethics consultants and ethics committees, find it moral to withdraw fluids and nutrition from dying patients at their advanced request, then this faith-experience should be trusted as contributing an important insight about the status of dying patients in technological environments.

There have been two responses, therefore, to moral challenge in the church in the past. The first has been to freeze the development of Christianity at that point and go no farther. Today's political fundamentalism seems to be characterized by this approach. From one point of view such moments are a failure of faith, a failure to believe that the Holy Spirit can redeem human society and human technology. Society and technology, too, are the "works of human hands" brought for transformation at the eucharistic table. Roman Catholics found they could not cope with historical thinking itself in the 19th Century. The Church condemned "modernism". All officials in the church, even deacons before becoming priests, were required to take an oath against modernism that was finally dropped in the waning days of Vatican II.

The other response to the disjunctive event is to grapple in some degree of darkness with the evils of the time. There is little question that American moral thinking is colored by our context too. There seems to be a demeaning of the value of human life and persons that occurs in our society. The values are skewed. Hence, the debate between American and Vatican bishops demonstrated not only disagreements about moral teaching, but more profoundly, disagreements about **methodology** to be used in the analysis, either by appeals to tradition, or by appeals to experience. This important point about which MacIntyre alerts us can be seen from the attack on and defense of American "values".

As the Vatican spokesmen at the meeting with American archbishops claimed, American society is sick. It is obsessed with sex. Cardinal Gagnon pointed to then-current TV shows such as "Dallas" as encouraging promiscuity, and Cardinal Silvestri claimed that this led to a high number of annulments in the American church. Both indicated that about 75 percent of American Catholics practice artificial birth control in defiance of church teaching [33]. They also implicated the "tender hearts" of women in annulment cases, betraying a sexism that would not be tolerated if it were voiced by any leaders in the United States.

Yet there is some truth to the claim. American compulsion about sexual attractiveness occurs at a time when laws are developing to properly respect the rights of women. In "Married With Children", once Fox Broadcasting's Archie Bunker type television program, its hero, raunchy Al Bundy, referred to women as "Dog doo-doo; The older they get, the easier they are to pick up" [14]. A couple years ago, Chief Justice Renquist held a rare news interview to appeal for greater salaries for justices in the United States. At that time, Vanna White on "Wheel of Fortune" reportedly drew down a greater salary than all nine Justices of the Supreme Court combined. This skewing of values has an impact on American children. A recent survey of 8th and 10th graders in the United States showed that one in three teens considered suicide; in 8th grade 77 percent had tried alcohol, rising to 89 percent by 10th grade; 34 percent of the students had reported being threatened by violence during the previous year, and 14 percent had been robbed. Tweny-three percent of the boys among the 11,000 students in 20 states surveyed said they carried a knife; 19 percent of the girls reported that someone had tried to force them into sex, including 6 percent who said that it happened at school [3].

We are a violent, abusive society. If this is how our children are raised, we must be extremely cautious when appealing to our experience, even if infused with faith, as a major source for current and future moral thinking. We are infected by the violence of our times and its disrespect for authority. Yet a major source of religious analysis is the current experience of the church. There is official teaching that we ought to consult the "priesthood of the laity", based on the extensive theology of Ives Congar and Vatican II [16].

But this official teaching clashes with statements by Vatican representatives like Cardinal Gagnon, who linked homosexuality, the use of artificial birth control, increased promiscuity, teenage pregnancies, and sterilization, with increasing feminism in America which, in his words, "has a deleterious influence on the family" [7]. The condemnation of birth control, sterilization, and feminism in one breath seems not to take the informed conscience and good will of many people into account. Some moral problems are more serious than others. As Merton again notes, "One could certainly wish that the Catholic position on nuclear war was half as strict as the Catholic position on birth control" ([49], p. 349). There seems to be a failure of belief in the redeemed goodness of people. One gets the impression that God of history cannot handle things well enough on His own, and that individuals must get obsessed about holding the line against any changes in His stead.

If there is a disjunction today between our present and our past, what resources do we have to confront modern moral challenges in health care? If the past is problematic, where are we to turn? If we turn to the future, what kind of thinking must we adopt?" [62]

4. EXPERIENCE AND THE FUTURE

The past cannot be transplanted. That means that responsibility to the past is doubled, at the very least. Not only must we work harder at trying to reconstruct the past context in order to properly understand and listen to the concerns of fellow-pilgrims, but also we must be doubly cautious about applying their thinking to our present condition. Theological bioethics must deconstruct present experience from a faith stand-point since such experience may be infected by our cultural violence, just as during a divorce, persons must be aware of how anger may be affecting their judgments. This requirement cuts us off from the past in a way that previous moral thinkers in the church were not cut off from their past.

Without this hard work, an appeal to tradition in bioethics is like Bush hiding behind the flag in his 1988 presidential campaign. The rhetoric is there. The commitments that rhetoric evoke are not present. Just as the virtue of patriotism has changed in a post-Nazi, post-Stalinist, post-Watergate era, so too is the virtue of honoring the teaching authority of the church. We must still do so, but the content of that honor is different. The salute is the same, but the awareness of caution while saluting, the increased responsibility of the individual, is different.

Because of the disjunctive nature of our age, thinkers must analyze more directly the experience of the faith-community. But this experience is tempered by the "markers" of our age already noted. So how can a past resource function in the future-directed experience we must employ? I offer a brief analysis and then will give a few examples.

Experience is in the present. Experience can be backward-looking or forward-looking. If it is backward-looking, the present experience will be fit into past experience and past moral teaching. Take the problem of genetic control. If we can control our genes, then we can adopt a theory of moral responsibility from other past control issues, like controlling the environment, or controlling the workplace, or controlling the moment of death. A forward-looking experience will stress the virtue of hope, it will be optimistic, it will stress an "ought" morality, how things ideally ought to be, and it will encompass the "building up of the Body of Christ". Past ideas can be applied to future problems, but with the difference of a new moral hermeneutic that does not rely excessively on the past.

To the question, for example, of whether or not a person is committing suicide by requesting not to be prolonged on machines, retrospective analysis might examine the problems of personal autonomy, the prohibition against taking one's life, and the like. But a new prospective moral hermeneutic might focus on the more important question: Should any human life-form be subject to machines or the enthusiasms of others, especially in an age that automatically makes persons slaves of the machine? Against the "Power of Babel" we might carve out a new right of human beings never to have to depend upon their own creation without their consent, whether or not they are dying.

What I suggest here is that today's moral analysis requires a new moral viewpoint. This viewpoint, as James Gaffney argues, is often left unoccupied as we grapple with a moral issue. The viewpoint itself is required to even ask the moral question. Gaffney argues that most of us can make sense about moral judgments and have well-oiled moral principles. "We simply fail at times to assume any moral point of view", most often at times we are dealing with our own affairs ([28], p. 57). If we remain too immersed in current experience without any moral viewpoint, suffering lassitude from the imprecations of disjunctive experience like the citizens in *Clockwork Orange*, then no moral judgments will be made about the unconscionable dependency of the human spirit on objects, things, techniques of its own making. Every attempt at setting an ethical policy ought to take the future implications of that policy and the context in which it is formed into account.

How would those moral judgments differ from efforts to apply the past as if there were no disjunction? I will offer a few examples.

The first example might be the condemnation of *in vitro* fertilization in the Vatican's "Instruction" of 1987. In the "Instruction", *in vitro* fertilization, along with many other reproductive technologies, is condemned as morally wrong. This view is based on an application of a Natural Law Ethic almost at the very start of the document that logically leads to such condemnations. Even though one cannot state with certainty that embryos are human persons, nor has the church taught it, the document argues, it is safer to consider them thus, and ascribe to them the rights of all persons [17]. Once one accepts this principle, no "manipulation" of human persons can take place without consent. Since embryos cannot give consent, no manipulation, however rewarding (e.g., for childless couples) is permitted. Other problems with applying past teaching in the document also occur, e.g., the problem of interpreting the unitive and creative function of marital intercourse [48].

What is important to note for our purposes, however, is that the document is almost completely ahistorical in method. It applies the principle of human dignity to all forms of human life in a deductive fashion, not once acknowledging that such reasoning and the Natural Law theory itself have come under intense questioning in the past 100 years.

Furthermore, the reasoning does not take into account the vast and complex web of science and technology noted above. Bioethicist James Drane, a participant at a conference examining the document when it was first published (University of Dayton, October 29–31, 1987), wondered out loud if the Catholic Church were in danger of becoming the Mennonites of the 20th Century. Recent changes in our ability to clone or more accurately "twin" embryos have made the concerns of the document even more relevant while simultaneously leaving many of its conclusions in the dust [15], [2], [51].

If the genome project is successful, and we map all human genes within the next ten years, and then if we follow course, we invent ways to apply our ability to transplant genes or alter them, *in vitro* fertilization will be the most moral way to have children in the future. Those who have children the old-fashioned way will be playing roulette with their children's future, dealing them possible genetically-based diseases. I would not be surprised if society would not pay for any health care needs such children have, when a method existed to eradicate those diseases that the parents did not follow. Hence, while concerns expressed in the document about the value of human life and our respect for it are certainly valid, its method and its conclusions will be overturned by advancing technical capacity. A prospective hermeneutical analysis is badly needed in this most important moral area.

Consider, too, the problem of euthanasia. One feature of Catholic moral teaching is the prohibition against intending the death of another. A second feature is an apparently unwarranted assumption that the traditional role of the physician was to preserve life [1]. Not intending death is the reason that double-effect euthanasia can be permitted, namely, that one intends by increasing a dose of morphine, for example, to decrease the pain experienced by the dying person with the concomitant, but unwilled effect, of decreasing the respirations and causing the death [9]. This principle is clearly stated in the Ethical and Religious Directives: "It is not euthanasia to give a dying person sedatives and analgesics for the alleviation of pain when such measure is judged necessary, even though they may deprive the patient of the use of his reason, or shorten his life" ([76], n. 29).

Yet because we do such a poor job of controlling pain and personally addressing the suffering of dying persons in highly technical environments [72], the agonies of dying persons (including their dependencies on machines) have become a major motivating force for the development of a right to active euthanasia [80], [53]. It will no doubt be legal in the United States in 10 years [71]. Once it does become legal, another disjunction will occur. Unless Catholic-sponsored hospitals have something to offer in its place, we will look like they are in favor of pain and suffering during the dying process. Dying people will avoid Catholic institutions like the plague. Accordingly an initiative was sponsored by the Catholic Health Association to develop guidelines for caring for the dying.

What should be the alternative to active euthanasia? It should be a positive therapeutic plan constructed with the patient and for the patient that would bring about the best possible death. The principles of hospice, of personal extension to the dying, will be required. So, too, will be an effective pain control protocol, that would include double-effect euthanasia. The "comfort care" model would have to be anchored by a new will, an intent that sees death as a good for some people, and actively tries to carry it out. The past teaching prohibiting this intent makes no sense in a new technological environment that can close the door too often to death, only to prolong the suffering of human persons [78].

A third example of future-experience guiding health care ethics comes from Joseph Fuchs' reflections on the principle of gradualism. As he points out, this pastoral principle was developed largely in the Missions. By it is meant that even though a Divine Law is expressed, no one can expect an immediate jump from one form of living to another. Persons can only gradually change. The African chief with six wives cannot abandon them all at once, at the moment he becomes a Catholic. He can only "gradually" be expected to change his life-habits, as these are culturally ingrained.

Fuchs points out that other interpretations of the principle of gradualism include that the law itself is "gradual", since human articulation of the Divine Law is subject to change (Indeed there are conflicts in laws articulated by the Church that would, if interpreted strictly, force persons into sin). Furthermore, Fuchs himself holds that the nature of God is not to be so unbending either, so that even if a perfectly articulated law of God is presented, God himself might gradually change it [25]. This point is an important consideration about excessive legalism in church and health care conduct, and shows the thoroughly historical and evolving character of modern theological ethics.

This interpretative principle can be used in the dispute about promoting "safe sex" by recommending condoms for persons at risk of AIDS. Even though the current Catholic teaching against artificial contraceptives is clear, particular circumstances make it difficult for some people to implement that teaching. The principle also can be used in support of gradually "divorcing" an offending medical service (say, one that performs sterilizations) from a new corporate structure in a merger with a Catholic hospital. Note how these applications include a prospective interpretation of the principle than its original, missionary, development. In general medical ethics and hermeneutics are closely allied, and are even more so in theological bioethics [66], [67], [69], [20], [18].

5. CONCLUSION

A realistic theological bioethics is one that accepts the challenge of the violence and disjunctions of our time and does not shrink-wrap moral teaching into an idealized worship of past, culturally-conditioned statements. Accepting this view is to accept a prospective hermeneutical challenge when working in moral thinking. Some points to keep in mind are:

1) Morals develop just as doctrine does. A good example of thinking in this regard is the new applications of the principle of gradualism noted above.
2) A new moral hermeneutics is required. Up to now, we have applied principles and rules, interpretative rules such as the Double-Effect, Proportionality, and Gradualism itself, without sufficient attention to the future implications of the use of the rules, or even of the consequences of them for families and caregivers ([30], last chapter).
3) Humbly recognize the fallibility of the filter of our age, particularly when it is so violent and disjunctive. We cannot wish away this disjunction. Postmodern thinking must be incorporated into theological bioethics. As a consequence of it, we must learn to trust our current faith-community's experience in the moral realm in a way we did not have to in the past.
4) To maintain flexibility and adaptability in an ever-changing environment, some form of casuistry coupled with guiding norms is the best possible approach to the difficult problems we face [37]. Indeed, debates about the proper utilization of casuistry in Catholic bioethics continue [65], [79]. These will yield only contextualized results, not universally accepted principles.
5) Listening is essential for a moral hermeneutics of the present. Tolerance is more valuable than moral belligerence because the latter too closely identifies certain moral actions or inactions with the holiness and authority of God. Nonetheless, radical moral disagreement does occur, and spokespersons for churches have a right to try to influence public policy [12], [40]. The opposite is also true. Leaders in the church have a right to question public policy stands. The complaint of the German theologians

about the Pope's stand on artificial birth control and American advertise-
ments in the paper on this point are based in the reaction to method [5].
There are many important sides to each issue. To cut them off too early
by making pronouncements, or pushing for legislation, violates the principle of listening. The realm of morality is one that admits only of moral
certitude, as St. Thomas called it *ut in pluribus*, generally for the most
part true.
6) Finally, theological bioethics must keep in touch with the experience of
believers, especially about health care matters. Almost everyone has a story
to tell about their parents or grandparents. And these stories revolve around
the loss of freedom and dignity brought on by dependency on machines.
We thinkers do not shape belief. We interpret it. The believing community is the best test of our ideas.

Loyola University Chicago Stritch School of Medicine,
Maywood, Illinois, U.S.A.

BIBLIOGRAPHY

1. Amundsen, D.W.: 1989, 'The Physician's Obligation to Prolong Life: A Medical Duty
   Without Classical Roots', in Robert M. Veatch (ed.), *Cross Cultural Perspectives in Medical
   Ethics: Readings*, Jones and Bartlett, Boston, pp. 248–262.
2. Annas, G.J.: 1994, 'Regulatory Models for Human Embryo Cloning: The Free Market,
   Professional Guidelines, and Government Restriction', *Kennedy Institute of Ethics Journal*
   4(3), 235–250.
3. Anonymous: 1989 (March 10), '1 in 3 Teens Considered Suicide', *Chicago Tribune*, Sec.
   1, 16.
4. Anonymous: 1989 (March 17), 'Chileans Blame U.S.', *Chicago Tribune*, Sec. 1, 1.
5. Anonymous: 1989 (January 27), 'Professors Challenge Pope's Role', *Chicago Tribune*,
   Sec. 1, 5.
6. Anonymous: 1989 (March 9), 'U.S. Bishops in Rome', *Chicago Tribune*, Sec. 1, 1.
7. Anonymous: 1989 (March 11), 'Vatican Aide Condemns U.S. Morals', *Chicago Tribune*,
   Sec. 1, 3.
8. Anonymous: 1989 (March 17), 'Vatican Requires New Fidelity Oath for Church Officials',
   New Catholic Explorer, 3.
9. Ashley, B. and O'Rourke, K.: 1982, *Health Care Ethics*, 2nd Edition, Catholic Health
   Association, St. Louis, MO. Newest Edition is published by Georgetown University Press.
10. Baron, R. (ed.): 1951, Hugh of St. Victor, *Epitome Dindimi in Philosophiam*, *Traditio*,
    11, 151ff.
11. Bellow, S.: 1959, *Henderson, the Rain King*, Viking Press, New York.
12. Boyle, J.: 1994, 'Radical Moral Disagreement in Contemporary Health Care: A Roman
    Catholic Perspective', *Journal of Medicine and Philosophy* 19(2), 183–200.
13. Callahan, S.: 1989, 'Ethical Blocks and Biases', *Health Progress* 70(1), 20, 22.
14. Clark, K.R.: 1989 (March 12), 'Tigress of the Tube: Housewife Makes Every Letter Count
    in Attack', *Chicago Tribune*, Sec. 5, 2.
15. Cohen, C.B.: 1994, 'Future Directions for Human Cloning by Embryo Splitting: After the
    Hullabaloo', *Kennedy Institute of Ethics Journal* 4(3), 187–192.
16. Congar, I.: 1965, *Lay People in the Church*, 2nd Revised Edition, Newman Press,
    Westminster, MD.

17. Congregation for the Doctrine of the Faith: 1987 (March), *Instruction on Respect for Human Life in Its Origin and on the Dignity of Procreation*, Polyglot Press, Vatican City.
18. Cooper, H. Wayne: 1994, 'Is Medicine Hermeneutics All the Way Down?', *Theoretical Medicine* 15(2), 149–180.
19. Cox, H.: 1965, *The Secular City*, MacMillan, New York.
20. Daniel, S.L.: 1994, 'Hermeneutical Clinical Ethics: A Commentary', *Theoretical Medicine* 15(2), 133–140.
21. Dilthey, W.: 1962, *Pattern and Meaning in History*, Harper Torchbooks, New York.
22. Dyer, G.: 1989 (March 20), 'The Secularizing Evolution That Includes Islam', *Chicago Tribune*, Sec. 1, 13.
23. Engelhardt, H.T., Jr.: 1991, *Bioethics and Secular Humanism*, Trinity Press International, Philadelphia, PA.
24. Fletcher, J.: 1988, 'The Moral Limits of Knowledge', *The Virginia Quarterly Review* 64(4), 565–584.
25. Fuchs, J.: 1981, *Christian Morality: The Word Becomes Flesh*, Georgetown University Press, Washington, D.C.
26. Funk, R.W.: 1966, *Language, Hermeneutic, and Word of God*, Harper and Row, New York.
27. Gadamer, H.G.: 1965, 2nd Edition, *Wahrheit und Methode*, J.C.B. Mohr (Paul Siebeck), Tübingen.
28. Gaffney, J.: 1988 (January 23), 'Moral Views and Moral Viewpoints', *America* 158(3), 55–59, 76.
29. Gallagher, M.: 1989 (May 19), 'No Charges in Life-Support Case', *USA Today*, 3A.
30. Graber, G.C. and Thomasma, D.C.: 1989, *Theory and Practice in Medical Ethics*, Continuum Press, New York.
31. Heidegger, M.: 1966, *Discourse on Thinking*, Harper and Row, New York.
32. Hirsley, M.: 1989 (March 10), 'Blame U.S. Society, Bishops Tell Pope', *Chicago Tribune*, Sec. 1, 12.
33. Hirsley, M.: 1989 (March 17), 'Each Side Scored at Vatican Talks', *Chicago Tribune*, Sec. 2, 9.
34. Hirsley, M.: 1989 (March 12), 'U.S. Bishops Predict No Dramatic Changes', *Chicago Tribune*, Sec. 1, 3.
35. Hoffmann, R. Joseph: 1988, 'The Moral Rhetoric of the Gospels', in R. Joseph Hoffmann and Gerald A. Larue (eds.), *Biblical v. Secular Ethics: The Conflict*, Prometheus Books, Buffalo, NY., pp. 57–68.
36. Jameton, A.: 1995, 'Human Activity and Environment Ethics', in D. Thomasma and T. Kushner (eds.), *Birth to Death: Biology, Medicine, and Bioethics*, Cambridge University Press, Cambridge. In press.
37. Jonsen, A. and Toulmin, S.: 1988, *The Abuse of Casuistry: A History of Moral Reasoning*, University of California Press, Berkeley.
38. Jung, P.B.: 1988 (Fall), 'Abortion and Organ Donation: Christian Reflections on Bodily Life Support', *Journal of Religious Ethics* 16(2), 273–306.
39. Kass, L.: 1985, *Toward a More Natural Science*, The Free Press, New York.
40. Khushf, G.: 1994, 'Intolerant Tolerance', *Journal of Medicine and Philosophy* 19(2), 161–180.
41. Lobkowicz N.: 1967, *Theory and Practice*. University of Notre Dame Press, Notre Dame, IN.
42. Löwith, K.: 1949, *Meaning in History*, University of Chicago Phoenix Book, Chicago.
43. MacIntyre, A.: 1984, *After Virtue*, 2nd Edition, University of Notre Dame Press, Notre Dame, In.
44. MacIntyre, A.: 1990, *Three Rival Versions of Moral Enquiry: Encylcopedia, Genealogy, and Tradition*, University of Notre Dame Press, Notre Dame, IN.
45. Mandelbaum, M.: 1967, *The Problem of Historical Knowledge: An Answer to Relativism*, Harper Torchbooks, New York.

46. Marshall, P.M., Thomasma, D.C. and Bergsma, J.: 1994, 'Intercultural Reasoning: The Challenge for International Bioethics', *Cambridge Quarterly of Healthcare Ethics* 3(3), 321–328.
47. Masthene, E.G.: 1968 (July 12), 'How Technology Will Shape the Future', *Science* 161, 135–143.
48. McCormick, R.: 1987 (July/August), 'Document is Unpersuasive', *Health Progress* 68(6), 53–55.
49. Merton, T.: 1986, *The Hidden Ground of Love*, Farrar, Strauss, & Giroux, New York.
50. Merton, T.: 1966, *Raids on the Unspeakable*, New Directions, New York.
51. National Advisory Board on Ethics in Reproduction: 1994, 'Report on Human Cloning Through Embryo Splitting: An Amber Light', *Kennedy Institute of Ethics Journal* 4(3), 251–282.
52. Niebuhr, H. Richard: 1963, *The Responsible Self*, Harper and Row, New York.
53. Nowell-Smith, P.: 1987 (March), 'A Plea for Active Euthanasia', *Geriatric Nursing and Home Care*, 23–24.
54. Pellegrino, E.D. and Thomasma, D.C.: 1981, *A Philosophical Basis of Medical Practice*, Oxford University Press, New York.
55. Pellegrino, E.D. and Thomasma, D.C.: 1988, *For the Patient's Good: The Restoration of Beneficence in Health Care*, Oxford University Press, New York.
56. Pellegrino, E.D. and Thomasma, D.C.: 1993, *The Virtues in Medical Practice*, Oxford University Press, New York.
57. Pieper, J.: 1964, *Scholasticism: Personalities and Problems of Medieval Philosophy*, McGraw Hill Paperbacks, New York.
58. Rahner, K.: 1959, *Free Speech in the Church*, Sheed and Ward, New York.
59. Rahner, K.: 1968, *Hearers of the Word*, Herder and Herder, New York.
60. Rahner, K.: 1968, *Servants of the Lord*, Herder and Herder, New York.
61. Schillebeeckx, E.: 1951, *Christus: Sacrament van de Godsontmoeting*, H. Nelissen, Bilthoven.
62. Schillebeeckx, E.: 1968, *God, The Future of Man*, Sheed and Ward, New York.
63. Sharp, D.E.: 1943, 'The *De ortu scientiarum* of Robert Kilwardby', *The New Scholasticism* 8, 1–30.
64. Spoerl, J.S.: 1994 (Spring), 'Impartiality and the Great Commandment: A Reply to John Cottingham (and Others)', *The American Catholic Philosophical Quarterly* 68(4), 203–210.
65. Tallmon, J.M.: 1994, 'How Jonsen Really Views Casuistry: A Note on the Abuse of Fr. Wildes', *Journal of Medicine and Philosophy* 19(1), 103–114.
66. ten Have, H.: 1994, 'The Hyperreality of Clinical Ethics: A Unitary Theory and Hermeneutics', *Theoretical Medicine* 15(2), 113–132.
67. ten Have, H.: 1994, 'Philosophy of Medicine and Bioethics, World-Wide and Nation-Focused', *European Philosophy of Medicine and Health Care* 2(2), 3–5.
68. Thomasma, D.C.: 1983, *An Apology for the Value of Human Life*, Catholic Health Association, St. Louis, MO.
69. Thomasma, D.C.: 1994, 'Clinical Ethics as Medical Hermeneutics', *Theoretical Medicine* 15(2), 93–112.
70. Thomasma, D.C.: 1994, 'Moving Beyond Autonomy to the Person Coping With Illness', *Cambridge Quarterly of Healthcare Ethics* 3(4), forthcoming.
71. Thomasma, D.C.: 1988, 'The Range of Euthanasia', *Bulletin of the American College of Surgeons* 73(8), 4–13.
72. Thomasma, D.C. and Graber, G.C.: 1990, *Euthanasia: Toward an Ethical Social Policy*, Continuum Books, New York.
73. Thomasma, D.C., Muraskas, J., Marshall, P., Tomich, P. and Myers, T.: unpublished, 'What Did We Learn from the Lakeberg Twins?'
74. Tillich, P.: 1952, *The Courage to Be*, Yale University Press, New Haven, CT.
75. Tyson, R.: 1989 (March 31-April 2), 'Skipper Fired; Oil Spill Probe Cites Drinking', *U.S.A. Today*, 1A.

76. U.S. Catholic Conference: 1989, *Ethical and Religious Directives for Catholic Health Care Facilities*, Catholic Health Association, St. Lous, MO.
77. Weaver, M.A.: 1994 (September 12), 'A Fugitive from Injustice', *New Yorker* 70(28), 48–60.
78. Westley, R.: 1994, *Right To Die*, Twenty-Third Press, Mystic, CT.
79. Wildes, K.W.: 1994, 'Respondeo: Method and Content in Casuistry', *Journal of Medicine and Philosophy* 19(1), 115–120.
80. Wrable, J.: 1989 (January 20), 'Euthanasia Would Be a Humane Way to End Suffering', *American Medical News* 31.

ROBERT M. VEATCH

Theology and the Rawlsian Original Position: Inventing and Discovering Moral Principles

The social contract theory of John Rawls [9] involving the *original position* methodology has emerged as a major technique in secular philosophy (including secular bioethics) for establishing a set of basic moral principles. Disputes have arisen over how this type of social contract theory relates to traditional theological ethical systems such as those of the monotheistic religions of the Western world ([11], cf. [13]).

Here I want to argue that, properly understood, the secular-appearing social contract of the Rawlsian type is compatible with the more traditional theological ethics. To put my claim more specifically, I suggest that some metaethics that appear to be *inventing* norms through a social contract process, in fact, are *discovering* them and that the process of discovery is compatible with theological ethics that discover ethical norms.

To be sure, some religious ethical systems that discover norms use epistemologies that would be incompatible with social contract theory. For instance, an epistemology that relies on divine revelation while stressing human fallibility and the inability to know through reason would be very different from and incompatible with social contract theory that relies on human reason, but these would only be epistemological differences, not necessarily differences in the substance or content of the norms. They would not be differences regarding the claim that the ethical principles or norms are conceptualized as preexisting the human process that leads to articulating them. In this essay, I would like to explore the relation between the Rawlsian original position and theological ethics. I will argue that some secular social contract ethics do, in fact, act as if the contractors were *inventing* a set of principles for a society while others, including the Rawlsian social contract, are more appropriately seen as *discovering* them and that the secular discovering theories are compatible with (and perhaps dependent upon) certain theological or quasi-theological methods for establishing ethical principles.

75

E. E. Shelp (ed.), Secular Bioethics in Theological Perspective, 75–84.
© *1996 Kluwer Academic Publishers. Printed in the Netherlands.*

RAWLS AND THEOLOGY

Rawls's Original Position Social Contract

Rawls addresses the problem of what it means for someone to claim that a particular ethical principle ought to govern basic social practices and how humans may know those principles. He claims that a set of ethical principles is more reasonable or more justifiable if "rational persons in the initial situation would choose its principles over" the alternatives ([9], p. 17). The concept of the initial situation, or what he comes to call the *original position*, is crucial to his theory. Its meaning and function are well analyzed in the philosophical literature. In simple terms, the original position is a purely hypothetical construct in which one imagines persons coming together to make some fundamental choices about the moral structure of a society. Rawls discusses at length what the characteristics of these persons would have to be in order for the result to be considered a set of *ethical* principles. He envisions them to be heads of households who will articulate the "basic structure of society" by creating a list of principles. They may be persons at any time in history (during the age of reason), having to make choices under conditions of moderate scarcity. They will operate under the formal constraints of "generality, universality, publicity, ordering, and finality" and will manifest "mutual disinterestedness" (limited altruism) and rationality.

These features are subject to dispute. For instance, recent feminist critics have attacked the notion that it is heads of households who would be envisioned as choosing the basic social principles ([1], p. 51). The key feature, however, is that all participants, no matter what their precise characteristics, would operate under a *veil of ignorance*, under which they would know the general facts about human society – the general laws of science and psychology – for example, but they would not know any particulars about their places in society, class position or social status, fortune in the distribution of natural assets and abilities, intelligence and strength, and the like ([9], p. 137).

The function of the fiction of a veil of ignorance is to make us imagine humans who consider equally the interests of all possible persons, avoiding choices that we can expect persons with our own special characteristics would favor.

The Implications of Rawls Social Contract for Theological Ethics

Many people working in theological ethics are distressed by the original position/veil of ignorance social contract approach. For some, the metaphor of contract is too legalistic, too businesslike ([8], [7], cf. [15]). That is an unfair criticism, however, since not all contracts are cast in the mold of a legal or business relation. The notion of a social contract has a long history in political philosophy having nothing to do with law or business. Religious ethics has accepted the contract metaphor in such notions as the "marriage contract".

Covenant theory is appropriately seen as a subtype of social contract. The image of contract per se need not distress theological ethicists.

The more troublesome aspect of the social contract metaphor is that, to many, it implies human beings coming together to create a set of agreements that will henceforth be binding on a community. Regardless of whether this agreement is interpreted legalistically, it is seen as humans *inventing* the social norms. That surely is contrary to a theological metaethics – at least for the monotheistic world that holds that a divine power is the creator of the universe (and therefore of the ethical structure of the universe). Ethics is, from this point of view, not a matter of mere human invention.

Social Contract and Inventing Norms

Some versions of secular social contract theory clearly convey real-life human beings inventing sets of norms for social groups. These norms, once invented, are taken to be binding ([3], p. 41). The view in which moral principles are invented raises some serious theoretical problems such as why one would feel obliged to keep to an agreement when it was not in one's interest to do so. One is tempted to say that it is because one has promised to keep the social contract. That, however, simply raises the question of why one is obligated to keep one's promises unless there is some preexisting moral obligation to do so. Of course, it is those moral obligations that holders of this view say are invented.

Often, it comes down to the claim that it serves not only the general interests of the community, but the interests of each individual, to keep commitments once they are made. That claim seems highly implausible empirically. Surely, there are cases in which it is contrary to one's interest to keep an agreement even after one factors in the disutilities of gaining a reputation of a contract-breaker if one is caught.

More fundamentally, however, it is not clear why contracts of this sort between real people made out of perceived long-term mutual self-interest in maintaining a peaceable community would count as *ethics*. Even if such agreements count as prudent self-interest, they hardly manifest the characteristics we normally associate with ethics. Ethics usually requires such characteristics as universalizability, generality, and altruism (or at least neutralism). It also usually requires that the criterion of publicity be satisfied, which holds that people must be willing to state publicly their moral positions.

Assuming one could figure out why contracts ought to be kept, inventing norms using actual social contracts could create mutual sets of obligations for social groups, but presumably different groups could come together to invent different sets of norms. Since ethics has no ultimate or universal frame of reference, different groups would be free to invent different norms. This is what some call social relativism; others would consider it not ethics at all, but merely a set of social preferences.

Discovering Moral Norms Using a Hypothetical Social Contract

There is no reason to believe that Rawls or his followers had anything like this in mind. Their social contract is much more abstract. It formulates a set of principles or norms that are dictated by reason – to which all rational persons ought to assent. They would assent – by definition – if they were not biased by knowledge of their specific position in the social system. Rawls ([9], p. 138) holds that in the original position "rational deliberation satisfying certain conditions and restrictions would reach a certain conclusion" about the moral principles. Rawls holds that all rational persons who understand the concept of the veil of ignorance and are not biased by their unique position in the world would reach the conclusion that favors, for example, the principles of justice over the principle of utility. "The restrictions [in the original position] must be such that the same principles are always chosen"([9], p. 139).

If that is true, then it really does not make sense to say that the contractors in the original position invent anything. They appear compelled to it. Presumably, that must mean that something doing the compelling preexists the social contract. The contract is more of a "compact", a formalized summary articulating agreement about the existence of preexisting moral norms. In Rawls's case, the norms that preexist are the principles of justice that include equality of liberty, fair equality of opportunity, and the difference principle. These take priority over the utilitarian's principle of maximizing the aggregate good in a community (even if that comes at the expense of certain individuals in that society who may be very poorly off). This looks very much like the hypothetical contractors discover the norms; they are forced to them by some conditions of the world that preexist the contractors' agreement.

What Preexists the Social Contract?

For Rawls what preexists the contract is apparently rationality. It attaches to all humans who are in the original position. Rawls also holds that certain features of human psychology are also universal among those in the original position. For example, they have a "sense of justice" that allows them to judge whether "things are just or unjust" and to support these judgments with reasons ([9], p. 46). Other contractarian theories, such as those of Locke and Hobbes, might make other presuppositions about pre-contract human psychology. The point is that there exists a universe that shapes how all persons would think when they reflect on what the moral norms ought to be.

Empiricists, such as the Scottish moralists, held that there was a "moral sense" that was part of human psychology; natural law theorists, that natural laws or a human telos existed that could provide such a framework.

For a wide range of naturalist metaethics, there is something in nature that provides a foundation for morality. While there is some tendency to differentiate the contract theorists from the naturalist empiricists, rationalists, and natural law theorists, those contract theorists of the Rawlsian variety who hold that there is some force that will drive contractors to accept a certain set of principles need not be seen as opposing these theorists. Hence, Rawls is a rationalist contractarian; Locke an empiricist contractarian; and so forth. There is no reason a natural law theorist could not also be a contractarian provided the role of the contract is to articulate the moral structure that is part of the world order before the meeting of the contractors. Locke, for example, was a natural law theorist and a proponent of a social contract theory. Rawls's social contract is essentially an epistemological metaphor for discovering the moral reality. It says nothing about the ontological grounding of that reality or how that reality came about.

The Task of the Contractors: Articulating a Consensus

The key is what we expect contractors to do who are of the Rawlsian school, which is compelled by some force such as reason (or human psychology or the moral laws of nature) to articulate certain principles.

An analogy might prove useful. Consider an international conclave of the world's leading scientists called together to resolve an international dispute. We can imagine that scientists of the world working on some major scientific problem (such as the theory of relativity or the structure of the human genome) might attempt to reach a consensus by calling a worldwide conclave of the leading experts in the area under dispute. They might hold disparate positions about aspects of the matter under dispute.

The organizers could plausibly say to them that they realize that scientists from different countries, in different age and gender groups, and in different religious traditions may feel compelled to see the matters under dispute differently; all humans are, to some extent, biased by their sociocultural surroundings. The organizers might say that, for the purposes of this gathering, the scientific community will take as the definitive description of the matter under dispute any agreement that those attending would reach. The organizers might then say that as people come to the meeting they should try their best to put aside their preconceived notions and biases, adopting a "veil of ignorance" position.

Were a consensus statement to emerge from the meeting, we would not be inclined to say that the scientists *invented* the theory of relativity or the structure of the genome; we would say they articulated in a formal way what they had discovered. Moreover, they ought to acknowledge that their statement was not infallible. In some sense all such scientific theories are constructions of social groups. In this manner accounts of reality are socially constructed [2]. Nevertheless, to the extent that they were successful in assuming the veil of ignorance and thus blotting out their preconceived biases, the account they would give would be at least one of the possible descriptions of a preexisting reality. Since the world is infinitely complex and descriptions could be infinitely long, all accounts of the world are, to this extent, seen from the point of one world view [6]. It would be a "true" account in the sense that elements of the account could theoretically be falsifiable using empirical and/or rational methods.

We could say of such a scientific meeting that the goal was to set aside all idiosyncratic socio-cultural biases and come up with a socially constructed account of a preexisting reality that was one of the class of apparently truthful accounts. To the extent they succeeded, they would have *discovered* one of the *correct* accounts of the phenomenon they were assessing.

The Implications for Social Contract Theory

By analogy we might say that the hypothetical social contract theorists have as their goal *discovering* a *correct* account of the moral universe. They would acknowledge human fallibility, but to the extent that they succeeded in assuming the veil of ignorance and thus eliminated personal idiosyncratic biases, they would have provided one possible correct account of the moral principles. They would merely articulate their discovery, compelled to their conclusions by reason, human psychology, or the pre-existing moral structure of the universe. What they articulated would be socially constructed, but it would be one among many possible true accounts of the moral reality.

This raises the complex question of whether discovering and articulating a principle is different than interpreting or applying it to specific questions and whether discoveries and applications change over time. On the one hand, according to this view, to the extent that the account is a correct account, it is correct for all time or for at least a very wide range of possible worlds. Thus if hypothetical contractors behind a veil of ignorance would discover a principle of justice, that principle would be a real component of moral theory for at least a very wide range of worlds. Insofar as it is a principle of justice for distributing scarce resources, it presumably would not apply to some hypothetical world that lacked scarcity, but presumably most conceivable worlds have scarcity and the principle would apply. Many articulations of a discovered principle of justice may be shaped by the assumptions about the world in which one does the discovering. For example, Rawls at times claims that his principles are relevant to worlds of moderate scarcity. A fuller statement of the principles might be necessary for worlds of extreme scarcity (such as that of the Ik of Africa). It is not clear that the longer, more complete form is a "better" or "more accurate" statement, however. Newtonian physics is thought to be a correct account for some projects. Stating a full Einsteinian theory of relativity might actually be a worse, more confusing, and less relevant statement for those projects. But if one acknowledges the limits placed on any statements of a moral principle or a theory in physics, it still makes sense to say something like "To the extent the contractors have stated it correctly, they have stated it correctly for all times". Additional specifications may be necessary to clarify application to unusual environments, but insofar as the specifications or qualifications do not apply, the principle would apply to all worlds [10]. This is thus a realist metaphysics, although it is a nuanced one.

There is a second concern raised by the question of whether principles discovered using a contract method are the same for all times or whether they change over time. Contemporary philosophy of science and linguistic analysis makes clear that concepts and the meaning of language change over time. Different people in different settings may use the same terms in different ways. Thus certainly the same statements in ethics made in Biblical times and in twentieth century United States will have different nuances of meaning. Some of the differences arise from changes in meaning with the translation of terms from one language to another. Others occur when terms within a language change their meaning over the years. Nevertheless, we can presume some constancy of core meanings with at least some limited capacity to provide at least partial translation. To the extent that speakers of different languages or in different eras mean the same thing by their terms, it makes sense to speak of discoveries being correct and, if they are correct, they constitute one possible formulation that is correct for all times.

Even if we acknowledge that there is a partial capacity to provide translation from one linguistic and conceptual scheme to another, it does not follow that the moral principle discovered by those approximating the veil of ignorance is the only possible account that could have been discovered. Ideal observers might have looked from a different perspective and provided a different, though not incompatible, account. Thus two different social contracts could describe different aspects of the moral universe seen from different perspectives. In fact, to the extent that the observers stand in partially incommensurable worlds, they would have to provide at least partially different accounts. Nevertheless, for the realist philosopher and the monotheist theologian these accounts cannot be contradictory, once one allows for the differences in conceptual and linguistic usage.

THE IMPLICATIONS FOR THEOLOGICAL ETHICS

If this is a proper account of contemporary social contract theory of the hypothetical or ideal sort that Rawls has stimulated, what are the implications for theological ethics? Social contract theory is a kind of epistemological metaphor for what we mean by the claim we have an ethical principle and how we would go about knowing we had one. An ethical principle is a principle about ethical right-making characteristics of actions or rules that a group of rational, dispassionate persons would discover if they looked at the moral universe from behind the veil of ignorance. To the extent that real humans have successfully adopted that stance, they would be able to articulate an account of the principles they have discovered to be preexisting in the universe as required by reason, human psychology, and/or the moral laws of nature.

Secular philosophical ethics is thus divided over whether ethical principles are discovered or invented. Several major traditions in Western philosophical ethics share with theological ethics of the monotheistic sort the commitment to the view that principles are discovered to be preexisting in the laws of reason, human psychology, or the laws of nature. They would differ from the branch of theological ethics that is polytheistic; that holds to multiple gods with multiple ultimate moral authorities. They would also differ from the branch of monotheistic theological ethics that accepts a single foundation of morality, but relies solely on revelation as a means of human knowledge. It is not surprising, however, to find for every major naturalist metaethic a parallel theological metaethic: hence there are religious rationalist ethicists [4] just as there are Kantian and Rawlsian rationalist philosophical ethicists; there are religious empiricist ethicists [5] just as there are Humean secular empiricists; and there are religious natural law theorists [12] just as there are secular natural law theorists. The religious ethicists in these camps presumably would associate the grounding of ethical requiredness with a theological origin. How they know what the content of these religious norms is, is quite another matter. While some would surely insist on revelation, scriptural interpretation, or the teaching authority of the church, others might rely on reason or on more democratic consensus formation methods.

Protestant Christianity subscribes simultaneously to a doctrine of the priesthood of all believers and a strong doctrine of human fallibility. The priesthood of all believers affirms that each believer is an authority in interpreting the moral requirements of his or her religious tradition. The strong doctrine of human fallibility suggests that such interpretations are nevertheless subject to error. One approach in democratic polities is what can be called democratic centrism, in which the denomination collectively develops a moral consensus about the moral principles and their implications. It could be seen as adopting positions similar to the worldwide scientific meeting discussed above. It would not be stretching to say that they collectively articulate their understanding of the preexisting moral norms and that they would accept such Rawlsian notions as the veil of ignorance as a condition for proper discovery of the divine moral order.

Pre-Protestant social contract theory is more difficult to formulate. The medieval church relied on the teaching authority of the popes and church councils. Talmudic authority rested with rabbinical teachers and councils. Nevertheless, there is a Biblical basis for something akin to a social contract as a device for articulating the norms of a moral community.

In the book of Nehemiah, we are presented with what amounts to a social contract (or covenant) device for discovering and articulating the norms for the post-exilic community, which faced the problem of having lost its sense of the divine will. The text reports that "The rest of the people, the priests, the Levites, the gatekeepers, the singers, the temple, servants, and all who have separated themselves from the peoples of the lands to the law or God, their wives, their sons, their daughters, all who have knowledge and understand, join with their brethren, their nobles, and enter into a curse and an oath to walk in God's laws which was given to Moses the servant of God, to observe and do all the commandments of the Lord . . ." (Neh 10: 28–31). There follows, then, an articulation of the identified moral law that they articulate.

Clearly, having a group come together to jointly articulate what they have discovered is not incompatible with universalist, naturalist ethics – whether they be secular philosophical ethics or religious theological ones. Presumably, the theological theories will have a different account of how the moral norms that are discovered got there in the first place. They may have a much more ready-made answer to such a question. Theological creation doctrine provides a straightforward answer to a question that must be puzzling for secular philosophers. But the theological discoverers and the philosophical discoverers can easily share the same bed when it comes to a methodology for knowing what those norms are and articulating what is found to exist in the universe.

In fact, the theological ethicists may face an easier task than that of their philosophical brothers and sisters at certain points along the way. Theological ethics has no difficulties with the problem of how we know certain things that are necessary to get the system started. Theologians have a doctrine of faith that provides the initial premises – the presuppositions – that get a system started. By contrast, the philosophers invariably face an uncomfortable moment when they are pressed back to first premises or starting assumptions. They often feel compelled to cloak these in the disguise of respectability. Instead of simply stating initial beliefs or premises, they will often use phases such as "reasonable people will agree" or "reason shows that" or "it will be obvious that".

Rawls, himself, is often found positing certain premises that get his original position methodology going. For instance, Rawls says he *supposes* that "all have the same rights in the procedure for choosing principles; each can make proposals, submit reasons for their acceptance, and so on" ([9], p. 19). At one point he simply supposes that persons in the original position are equal, a presupposition that could end up affecting the relatively egalitarian principles of justice that finally emerge ([9], p. 19). In a previous treatment of these issues I suggested that Rawls may be "bootlegging" the Judeo-Christian tradition in these disguised faith statements ([14], p. 90). He assumes them, presumes them, supposes them, accepts them as premises, or posits that all reasonable people will agree with them. Without them his system will not go. They end up shaping his principles and making them reasonably compatible with Judeo-Christianity.

The bootlegging of assumptions or "faith moves" divorced from their institutional and cultural roots, however, is risky. I think it can be shown that Rawls ends up only a "fainthearted" egalitarian. He accepts inequalities (whenever they redound to the benefit of the least well off). He can see no reason for someone to prefer equality at a lower level of well-being when even the poorest might be a little better off if inequalities were permitted. He assumes anyone who would favor a more equal world in such cases manifests envy ([9], pp. 530–541). He cannot grasp the notion of moral solidarity and community that is so easily understood from within the more formal Judeo-Christian tradition.

CONCLUSION

Thus it is at least to some extent a mistake to see the Rawlsian tradition of hypothetical social contract to be at odds with monotheistic theological ethics. Some social contract theories clearly are in tension with any monotheistic religious point of view, especially the theories that rely on actual contractual agreements by actual social groups who feel no drive to universalize their conclusions. Some theological ethics are just as clearly in tension with hypothetical contact theories when the theological systems rely on more authoritarian epistemologies or on revelation. But democratic centrist theories of covenant in the religious traditions, which use the covenant as a way of articulating the human understanding of the moral law, need not be in any way incompatible with the hypothetical social contract theories that use contract as a metaphor for articulating the moral principles discovered through the use of rational reason, knowledge of human moral psychology, or the natural law.

Georgetown University
Washington, D.C. 20057

BIBLIOGRAPHY

1. Baier, A.C.: 1987, 'The Need for More than Justice', *Canadian Journal of Philosophy* 13 (Supplement), 41–56.
2. Berger, P.L. and Luckmann, T.: 1967, *The Social Construction of Reality*, Doubleday, New York.
3. Engelhardt, H.T.: 1986, *The Foundations of Bioethics*, Oxford University Press, New York.
4. Green, R.M.: 1978, *Religious Reason: The Rational and Moral Basis of Religious Belief*, Oxford University Press, New York.
5. Hutcheson, F.: 1971 [1728], *Illustrations on the Moral Sense*, The Belknap Press of Harvard University Press, Cambridge, MA.
6. Kuhn, T.S.: 1962, *The Structure of Scientific Revolutions*, University of Chicago Press, Chicago.
7. Master, R.D.: 1975, 'Is Contract an Adequate Basis for Medical Ethics', *Hastings Center Report* 5, 24–28.
8. May, W.F.: 1975, 'Code, Covenant, Contract, or Philanthropy?', *Hastings Center Report* 5, 29–38.
9. Rawls, J.: 1971, *A Theory of Justice*, Cambridge, Massachusetts, Harvard University Press.
10. Richardson, Henry S.: 1990, 'Specifying Norms as a Way to Resolve Concrete Ethical Problems', *Philosophy and Public Affairs* 19: 279–310.
11. Sumner, L.W.: 1982, 'Does Medical Ethics Have Its Own Theory?', *Hastings Center Report* 12, 38–39.
12. Thomas Aquinas: 1915, *Summa Theologica*, Fathers of the English Dominican Province (ed.), R & T Washbourne, Ltd., London.
13. Veatch, R.M.: 1981, *A Theory of Medical Ethics*, Basic Books, New York.
14. Veatch, Robert M.: 1986, *The Foundations of Justice: Why the Retarded and the Rest of Us Have Claims to Equality*, Oxford University Press, New York.
15. Veatch, R.M.: 1983, 'The Case for Contract in Medical Ethics', in E. Shelp (ed.), *The Clinical Encounter: The Moral Fabric of the Patient-Physician Relationship*, D. Reidel Publishing Co., Dordrecht, Holland, pp. 105–112.

KAREN LEBACQZ

The Weeping Womb: Why Beneficence Needs the still Small Voice of Compassion

THE CHALLENGE

At a meeting of the American Academy of Religion in 1991, Dr. William Bartholome, pediatrician and medical ethicist, argued flatly *against* the relevance of compassion as a guide for current medical practice. In our postmodern world, he suggested, autonomy has become the key to ethical behavior for the physician. Patients and physicians contract with each other for services. There is no room for anything beyond that contract and its specifics.

Whether Bartholome simply meant to be provocative or whether he was sighing in lament I do not know. I do know that his vision of modern medicine troubled me then and troubles me now. I have had experiences of genuine compassion in my health care. I believe that compassion is not only possible, but important as a base for medical ethics. As Beauchamp and Childress note, "in many contexts . . . the expression of compassion makes a critical moral difference" ([2], p. 467). I want compassion in my health care.

I suspect that Bartholome does as well. When he spoke of the lack of room for compassion, I think he was really speaking about beneficence and about its attendant problem: paternalism. "Doing good" for another runs the risk of imposing on that other my view of the good. It is this imposition of my will or perspective on another that Bartholome intended to exclude. Hence, he stressed contract with its implied limits to the imposition of one person's views on another.

A feminist theological understanding of compassion demonstrates precisely that compassion is not subject to the dangers of paternalism associated with beneficence. In this essay, I will consider briefly the understanding of beneficence in two contemporary secular theories of bioethics: the reigning autonomy paradigm offered by Beauchamp and Childress (and implicitly adopted by Bartholome) and the beneficence-based alternative proposed by Pellegrino and Thomasma. I will discuss the danger of paternalism in beneficence. I will then offer an abbreviated feminist theological view of compassion. Many dimensions of compassion will not be discussed here, as I will focus the discussion on the problem of paternalism. Ironically,

E. E. Shelp (ed.), Secular Bioethics in Theological Perspective, 85–96.
© 1996 *Kluwer Academic Publishers. Printed in the Netherlands.*

compassion gets short shrift both in traditional views of medical ethics and even in contemporary feminist views ([8], [24]). Yet it has distinct gifts to bring.

BENEFICENCE

Identified by the National Commission for the Protection of Human Subjects of Biomedical and Behavioral Research as one of the four basic principles that should underlie the conduct of research and the delivery of health care, the principle of beneficence has become cannonized in the popular textbook *Principles for Biomedical Ethics* by Beauchamp and Childresss. Beauchamp and Childress are skeptical about the range and scope of an obligation of beneficence ([2], pp. 264–265). In contrast to their skepticism, Pellegrino and Thomasma, in *For the Patient's Good: The Restoration of Beneficence in Health Care*, take beneficence to be the central and overriding principle for the ethical practice of medicine ([18], p. vii).

What is beneficence? Beauchamp and Childress define it as "all forms of action intended to benefit other persons" ([2], p. 260). Pellegrino and Thomasma define it in the medical context as "positive enhancement of all the components packed into the complex notion of the patient's good" ([18], p. vii). In shortest form, beneficence is simply doing good for another. As an ethical principle, beneficence expresses the notion that there is an obligation or duty to do good for the other.

Beauchamp and Childress propose that beneficence encompasses three interrelated duties: (1) the duty to do good or confer benefits; (2) the duty to prevent or remove harm; and (3) the duty to balance goods against harms ([2], p. 260). My focus is on the first of these: the duty to confer benefits on another.

While recognizing that we often do good as a supererogatory act, Beauchamp and Childress do find a limited scope for an *obligation* of beneficence: I have a general duty to do good where (a) someone is threatened with loss, (b) I can prevent that loss, (c) my action is likely to prevent the loss, (d) my action is not particularly risky to me, and (e) the benefit to the other outweighs the risk to me. Thus, for example, I have a duty to give blood where it is needed to save a life and where I have no disease that would make giving blood risky for me or for the other.

Beauchamp and Childress recognize that this general duty of beneficence might take more specific form in roles defined by institutions. This is precisely where Pellegrino and Thomasma enter. Physicians, they argue, have an obligation to act in the patient's interest even at some personal risk (hence, modifying criterion d above) ([18], p. 27). Medicine is "of necessity" a form of beneficence, they propose; it is a response to the need of a sick person for help ([18], p. 32). The patient's needs must be the primary concern of the physician. Where Beauchamp and Childress are somewhat cautious regarding an obligation of beneficence, then, Pellegrino and Thomasma take beneficence to be the *sine qua non* of medicine and medical ethics.

PATERNALISM

Entire ethical systems such as utilitarianism have been built on the idea that we have a fundamental moral obligation to do good in the world, and to ensure that the good outweighs the harm. The Golden Rule – do unto others as you would have them do unto you – is often taken as a basic statement of the requirement of beneficence. Why, then, is there any problem with recognizing beneficence as a fundamental moral obligation?

Beauchamp and Childress find a fundamental danger in adhering to a principle of beneficence. In conferring benefits on another, we run the risk of imposing our own views on that other and forcing the other to do what we wish rather than what s/he wishes. This is paternalism: intentional overriding of a person's known preferences, where the one who overrides justifies that action by the goal of benefitting the one overridden ([2], p. 274). Hence, for any ethical system that holds personal autonomy as a central principle, it appears that beneficence runs the danger of violating the autonomy of another. "Weak" paternalism allows overriding the other's wishes only where the other's autonomy is presumed to be diminished; "strong" paternalism allows overriding the other's wishes even where the other is competent and autonomous ([2], p. 277).

Does this danger emerge in a system based on beneficence, such as that proposed by Pellegrino and Thomasma? Pellegrino and Thomasma speak repeatedly of a consensus between patient and physician, and urge that "there is to be no imposition of values, or decisions made in the best interests of patients without their participation" ([18], p. 33). In theory, therefore, there is no danger of strong paternalism and only weak paternalism is allowed: "Unless the patient is incompetent", they argue, "the physician is obligated to act for the good conceived by the patient and to support the patient's goals" ([18], p. 76).[1]

Yet Pellegrino and Thomasma also say that "both autonomy and paternalism are superseded by the obligation to act beneficently" ([18], p. 32). Further, they set limits on autonomy, suggesting for example that it is wrongly exercised if a patient rejects penicillin treatment for pneumococcal meningitis ([18], p. 47). This gives the impression that beneficence trumps autonomy and that the physician could override the wishes of the patient for the patient's good as conceived by the physician. At least *some* strong paternalism appears to be encompassed.

The danger that Bartholome notes is a real danger: under the guise of doing good, physicians might impose their own definitions of reality on patients. In our post-modern world such imposition of values is not acceptable. We retreat to autonomy and contract in order to preserve the freedom of each individual. It is no wonder that Bartholome rejected what he called "compassion" if by that he really meant "beneficence" with its attendant dangers of paternalism.

But do these same dangers of paternalism arise when we consider compassion?

COMPASSION

By contrast with beneficence, compassion is not simply doing good. It is "that feeling or emotion experienced when one is *moved* by the suffering or distress of another" ([23], p. 206). Prosaicly expressed in Wester's *New Collegiate Dictionary* ([30], p. 229), compassion is "sympathetic consciousness of others' distress together with a desire to alleviate it"; or, as Wendy Farley puts it in her study of *Tragic Vision and Divine Compassion*, it is "a mode of relationship and a power that is wounded by the suffering of others and that is propelled into action on their behalf" ([5], p. 69).

These definitions suggest that there are two aspects to compassion: first, "suffering with" the other (compassion comes from the Latin *com patior*, which means to suffer with or to experience with ([19], p. 84), and second, being impelled to counteract the suffering. Compassion includes the effort to alleviate suffering or remove or prevent harm to the other; in this, it is similar to beneficence. But it also includes a personal response of being moved. Compassion is not knowing *about* the pain and suffering of others. It is entering *into* that pain and suffering, "sharing it and tasting it insofar as that is possible" ([6], p. 21). Compassion is passionate. It is this dimension of personal response that leads compassion to be understood largely as a matter of character or virtue rather than of principle or obligation.

It is also the dimension of personal response that I believe can keep compassion from falling into the trap of paternalism. A view of compassion informed by feminist theology lifts up dimensions that offer protections against paternalistic imposition of values.

MOVEMENTS OF THE WOMB

Matthew Fox and C.S. Song speak of compassion as a movement of the bowels or heart ([6], p. 20; [27], p. 163). Here, they draw on Greek tradition, in which the term "splangchna", referring to the "noble viscera" – the heart, lungs, spleen, liver – is translated "compassion" (cf [6], p. 28). But compassion is not simply a movement of the bowels or heart. The Hebrew term translated "compassion" is "rachamiem", or "movements of the womb" ([9], p. 66). Compassion is a movement of the womb. It evokes female imagery. When Isaiah talks about God's compassion – e.g. Is 54:10 – he is using female imagery. As Trible says, "in biblical tradition an organ unique to the female becomes a vehicle pointing to the compassion of God" ([28], p. 38). The semantic movement from the wombs of women to the compassion of God is a persistent and powerful metaphor in biblical faith ([28], p. 56).

Is there something important about this rooting of compassion in an image of women, women's wombs, and their implied link with mothering? I think that the "womb" imagery is important. Consider three functions or movements of the womb:

1. The womb "receives" the other. The fertilized ovum travels down the fallopian tube to implant in the womb. The womb is receptive. In a moment, I will try to specify why I think this is important for understanding compassion as opposed to beneficence.
2. The womb does not only receive. It also shapes and nurtures, nourishing growth and life. It provides "food" for the journey into new life.
3. Finally, the womb expels and thereby thrusts the other into new life, receiving at the same time a new shape itself.

The womb receives, nourishes, and expels. All three of these moments or movements may be significant for understanding the role of compassion.

COMPASSION AS RESPONSE TO SUFFERING

If compassion is a response to others' suffering, then it is best understood by attending to suffering. Ironically, little has been written on suffering, in spite of its centrality in health care.

Eric Cassell suggests that suffering is caused by a fundamental threat or injury to the personhood of the sufferer ([4], Chapter 3). Personhood is made up of a number of dimensions, and so suffering is a complex phenomenon. While physical pain can be a cause of suffering, it is so largely when that pain impinges on other dimensions of personhood. What causes a person to suffer is unique and distinctive to that person, and can be understood, suggests Cassell, only by asking that person: "the only way to learn whether suffering is present is to ask the sufferer" ([4], p. 44).

Where Cassell identifies the *cause* of suffering in threat or injury to personhood, Dorothee Soelle and Warren Reich identify *modes* or moments of suffering and hence of compassionate response. From Soelle's study of suffering [26], Reich [19] draws three moments of suffering:

1. The first moment or dimension of suffering is mute suffering. "Mute" here does not necessarily mean unable to speak, but unable to express the experience of suffering. The person may scream but still not communicate to others the depths of the experience. In this phase, there is a significant loss of autonomy (cf. [22]).
2. The second moment of suffering is expressive suffering. Here, the sufferer begins to seek a language to express suffering. Often the first language or voice found is that of lament.[2] Sufferers turn to metaphors such as "victim" or "survivor".

3. The third moment is the fashioning of a new identity. Here, the sufferer finds a voice of her/his own. S/he begins to identify with the narrative created through conversation. S/he not only tells her/his story but finds an interpretation of the suffering.

Reich ([19], pp. 93–98) has appropriated this three-fold understanding of suffering to propose that there are three corresponding movements of compassion. Each of these corresponds roughly to one of the movements of the womb:

1. *Silent empathy*: Silence here is a sign of respect. The listener tries to place herself beside the sufferer, to suffer with the other. Silence is important in allowing the sufferer to find her or his own voice and not to impose an alien description. Just as the womb begins with receptivity of the reality of the other, so the compassionate listener must begin with "silent empathy" with a willingness to receive and "hear the other into speech" (cf. [15], p. 202).

2. *Expressive compassion*: If the sufferer is to move from mute suffering to expressive suffering, the listener also has a role to play in offering language, images, metaphors. Naming gives power over what is named; thus, Reich proposes that even a dry, scientific or medical diagnosis can help to alleviate suffering because it names the problem and allows the sufferer to gain power through the naming. Sharing of one's own story of suffering can also be important. Just as the womb provides context and "food" for the journey of growth of the other, so the compassionate one provides context, metaphor, "food" for the journey of the other.

3. *New identity*: Just as the sufferer moves to a new identity, so does the listener. Making oneself vulnerable to the suffering of another means that we will be changed. In this way, compassion is not only "suffering with" or experiencing with but is also a change in the compassionate one. Just as the womb expels the other into life and in so doing receives a new shape itself, so the compassionate one will be changed at the same time that s/he births new identity for the sufferer.

The three moments of compassion identified by Reich as matching Soelle's dimensions of suffering can be seen as reflections of the activity of the womb: receptivity, gathering and nurturing, expulsion into new life. Reich emphasizes that these movements of compassion are not prescriptive. They are a description of various aspects of compassion that match the dimensions of suffering elucidated by Soelle. They also reflect the movements of the womb.

There is not space here to explore all three of these dimensions as they relate to the imagery of the womb. I will explore the first one – the mode of receptivity – in order to indicate how the shift to feminine imagery might be important and how compassion might avoid certain problems of paternalism.

RECEPTIVITY VS. PROJECTION

Hobbes saw pity (his cognate for compassion, although some contemporary scholars are very careful to distinguish the two) as the "imagination" or "fiction" of future calamity to ourselves ([10], p. 58). In other words, what we feel for the other when we feel pity (or compassion) is not so much the *other's* pain, as it is a *projection* of the imagined pain that we would feel if the misfortune of the other were to happen to us. It is this projection or imagination that allows pity.

By contrast, feminist philosophers and theologians suggest that compassion is not a projection, but a reception of the other. When an infant cries, suggests Nel Noddings, the mother comforts before she begins to analyze what is wrong ([16], p. 31).[3] The "logic" of the caring/mothering relationship is that care and relationship come first, and abstract problem solving comes later. There is a motivational shift toward the child, and this motivational shift for Noddings is the heart of the ethical enterprise.[4] The mother *hears*, receiving the distress of the other, and *acts*. There is no need to imagine or project "what would be my pain if I were in the other's place?" In contrast to Hobbes' definitions of empathy as "projecting" oneself out toward the other, Noddings argues that caring is not projecting myself but *receiving* the other ([16], p. 30).

A similar view is expressed by Wendy Farley in her theological reflections on compassion. Farley suggests that compassion begins with sympathetic knowledge, with the ability to have some knowledge of the other's suffering. Like Noddings, Farley sees this as direct knowledge of the other's suffering. It is "authentic, direct apprehension of another's situation" ([5], p. 71).

A more ambiguous and nuanced view is offered by Indian poet Naris Basu ([13], pp. 36–37):

My mother wept in silence
As only mothers do;
Pitying my misfortune
That she had suffered too.

My mother shared my discomfort
As only mothers do;
Pitying my misfortune
For she was suffering too.

My mother knew my anguish
As only mothers do;
pitying my misfortune
For she had lived it too.[5]

Here Basu suggests that there is something special about a mother's capacity to feel her child's pain. She weeps, she shares discomfort, she knows the child's anguish "as only mothers do". The mother suffers with, experiences compassion, "as only mothers do".

Yet there is an ambiguity in Basu's poem about mothers and daughters. The poem indicates that the mother responds as she does because "she had lived it too". Does the mother "share discomfort" and "suffer too" and feel compassion because she receives the child's pain directly, as Noddings and Farley would have it? Or does she share discomfort and suffer and feel compassion because she remembers her own pains as a woman in a sexist society and projects those onto her daughter?

Does compassion mean direct apprehension of the other's pain or is it a remembering and projecting of one's own experiences? Or is it both? What does it take to participate in another's feelings? [29]

Pity or compassion may be easiest to feel where one has had a similar experience: "she had lived it too". Oliver Sacks, in his discussion of his broken leg, speaks of finding a physician who had also experienced a broken leg, and finally feeling understood ([21], p. 183).

But it is not necessary to have experienced a pain in order to receive the pain of another. My male gynecologist has no experience of the pains caused by some of the diagnostic tests he must run on my female organs. Yet my clenched fists, scrunched face, and stifled screams communicate all too well my pain, and I often find his face ashen and his eyes clouded – no doubt mirroring my own – at the end of a particularly difficult examination. Thus, it is not shared experience that is necessary for compassion, but simply the ability to receive the pain of the other into ourselves so that we experience it directly and immediately. Compassion, then, is not limited to those circumstances in which we have had a similar experience ourselves.

Here, Hume's claim that we cannot receive another's passion immediately must be recognized. We are sensible, suggests Hume, not of the passion directly, but only of its effects; from these we infer the passion. For Hume, the physician does not *feel* my pain directly, but *infers* it from the scrunched face and clenched fists ([2], p. 467). However, the feminist view allows the possibility of direct communication of pain.

Why does it matter whether we project ourselves into an imaginary future in order to apprehend the other's state or simply apprehend directly? Why is this seemingly miniscule philosophical point important?

Avoiding Paternalism

The reason Bartholome shied away from compassion was his fear that compassion may lend itself to paternalistic interventions that demolish the autonomy of the other. As noted above, I believe that it is not really *compassion* to which Bartholome objected. It is *paternalism*, interference with the other's life-goals and designs. What he was concerned to prevent is precisely the imposition of my values on another.

But would compassion allow such imposition? If compassion is not simply projecting out of my own experiences and values, but receiving into myself from the genuine pain of the other, then compassion avoids any such imposition because it begins with the genuine suffering of the concrete other. Beneficence is socially and culturally informed, and runs the risk of paternalism. As Cassell suggests, however, suffering is idiosyncratic, although socially and culturally informed. If compassion is therefore also idiosyncratic – feeling the other's suffering so intensely that I am motivated to alleviate it – then it runs less risk of simply being the projection of my own cultural values.

A feminist theological perspective teaches that we are connected: person to person, person to political, each to everything [7]. To perceive oneself as primarily connected, not primarily separate, may be a critical skill here: it allows us to experience not simply the "projection" of ourselves onto the other in pity or fear of a similar fate, but the "reception" – or better, the connection – of the other into ourselves so that there is direct apprehension of the other's pain. To be able to "suffer with" or experience with another, one has to be connected to that other.

THE WEEPING WOMB

To argue for the primacy of compassion is to see ourselves as *embodied* thinkers, ones whose wombs (or hearts) matter. In classical language, compassion is a virtue. It is not, therefore, on a par with beneficence, which is a principle. One can do good without being good. Does the feminist emphasis on compassion simply move us into the arena of virtue rather than principle or norm?

I think not. Compassion changes the way we perceive the world. Kwok Pui Lan tells a story of a little girl named Ah Ching living in a poor peasant family in China. Ah Ching awakens one night to find that her father is trying to suffocate her – he wants to have a son instead of a daughter. In terror, the little girl flees to her grandmother's house. The grandmother takes her in, comforts her, tucks her in bed – and then, in the middle of the night, with trembling hands, suffocates her.

Kwok Pui Lan introduces this story by saying, "I would like to tell you a story, a story that makes people cry". This story does indeed make me cry. Hearing it, I experience both revulsion and compassion. I want to rush in and rescue Ah Ching. I want to yell at the system that values male children and disvalues female children. I want to make it different. My heart aches for the little girl, for her terror, for her betrayal by those she trusted. My heart also aches for the grandmother, for her trembling hands and her heavy heart as she does what she knows she must do for the sake of her son. I feel compassion for the little girl Ah Ching. I also feel compassion for Ah Ching's father, for her grandmother, and for her mother, who is scarcely present in the story. I think it is impossible to hear this story without having our hearts torn.

I think it *should be* impossible to hear this story without having our hearts torn. Analytic reasoning can explain the features of Chinese society that made it crucial for this father to have a son and impossible to support the life of a daughter. Yet surely anyone who could hear this story and respond *only* with clear analytic reasoning is morally deficient. Apparently the father himself was not able to complete the act of killing; the grandmother, torn between her love of her grandchild and her love of her son, completes the act, but only with trembling hands. There is so much pain in this story that it is difficult not to cry. And so it is that Kwok Pui Lan entitles her essay "God weeps with our pain".

Compassion is a response to suffering. But this means that we must have the capacity to be moved by another's suffering.[6] Embodied thinking means the ability to be wounded, to feel the other's pain so that it makes us cry. In Noddings' language, we must have the capacity to receive the other.

This capacity of receptivity is central to wombs. When God's "womb" moves in compassion for Israel, this indicates a central characteristic of God in our lives: a disposition to hear and respond to cries of pain. The womb weeps. Compassion is taking the other and the other's pain into us so that their pain becomes ours and we weep together. Softness, vulnerability, receptivity, the ability to be hurt, attentiveness to the pains of others[7] – this, at a minimum, is the meaning of compassion as "rachamiem," movements of the womb.

Such compassion is a discipline. To this extent, then, it is indeed a virtue. But it not simply a trait of character. It is a way of thinking, of perceiving and interpreting the world. I am inclined to agree with Sara Ruddick that the most profound change we could make to counteract the oppressions of sexism is not the inclusion of men in the private sphere but rather the inclusion of this discipline of compassionate thinking in the public sphere.

THE PUBLIC SPHERE

Compassion will not solve all problems in health care ethics. It has limits. Many years ago, Margaret Adams warned against a "compassion trap" for women. She proposed that women entering the professions tended to stay locked into "caring" modes and roles [1]. Today, there is a movement among feminists critical of the "caring" approach to ethics. Most feminists today believe that we need both caring AND rationality, both virtues *and* principles [14]. Beneficence as one principle among others may still be needed.

But compassion may also be needed. Where beneficence runs the risk of paternalism, compassion requires not the imposition of my will on another but the reception of that other into the very core of my being. The pain of another becomes my pain. The womb is moved. It weeps. We suffer together. Because we suffer together, our entire outlook is changed. When our outlook is changed, the world is changed. The fundamental view that "we are in this together" may be the most important change needed in a pluralistic and divided world. Compassion, therefore, is not a private feeling but is, as Farley put it, a power – a power to receive, to be changed, and to change the world. It is a political act that goes beyond beneficence because it has the capacity to break the boundaries of culture and society. By receiving the other, attending to the pain of the other, the compassionate one avoids the trap of paternalism and the limits of simple beneficence.

McGill University

NOTES

1. This "modern" view of the patient's good, they propose, is "in accordance with autonomy". Beauchamp and Childress accuse this view of being nothing but a "dressed up defense of the autonomy model": while beneficence may provide the rationale for health care, the actual boundaries of practice appear to be set by respect for autonomy ([2], p. 272).
2. Similarly, Brueggemann suggests that when the muted ones in Israel begin to speak, the voice or psalm of lament is often the first voice raised ([3], pp. 51–52).
3. I agree with Sara Ruddick that while some features of the mothering experience may be almost invariant and others, though changeable, may be nearly universal, nonetheless it is impossible to express those features without importing other features specific to class, ethnic group, and sex-gender systems ([20], pp. 340–351). Thus, there is no simple "phenomenology" of mothering apart from history and culture, as Noddings appears to desire.
4. In fact, for Noddings, ethics has to do not with problem solving or with *justification* for action, but with *motivation* for action ([16], p. 95).
5. In between the choruses quoted here are numerous other "voices" that speak.
6. In his discussion of the compassion of the "good samaritan", Kosuke Koyama notes that the other two had "passed by" ([11], p. 651). Compassion is the opposite of "passing by". Koyama also notes the similarities between this parable and the story of the young Prince Siddhartha, the future Buddha, who could not pass by when he saw a sick man. Compassion is the opposite of having one's "heart" hardened. It is attending to the suffering of the other.
7. Space does not permit here the development of this theme of attending. Attending includes intellectual capacities of judgment as well as the training of perception to notice elements which carry meaning ([25], pp. 280–295). An intriguing discussion of maternal practice as a blend of intellectual activities and disciplines of feeling that "attend to" the child is offered by Sara Ruddick in her reflections on maternal thinking. Ruddick argues that maternal thinking exhibits "attentive love" ([20], p. 347). And she notes that both Iris Murdoch and Simone Weil – neither of them mothers "nor especially concerned with mothers" – lifted up the value of attentive love. Indeed, Weil offers the image of the Grail which will belong to the first comer who asks the guardian of the vessel "what are you going through?" ([20], p. 348) To ask "what are you going through" is something only one who attends to the other can do. It is foundational to hearing the cries of the other, and therefore to responding in compassion. As Farley suggests, compassion is not simply a feeling, but an enduring disposition – a tendency to perceive the world with attention to pain and a readiness to respond to it ([5], p. 74).

BIBLIOGRAPHY

1. Adams, Margaret: 1971, 'The Compassion Trap', in Vivian Gornick and Barbara K. Moran (eds.), *Woman in Sexist Society: Studies in Power and Powerlessness*, New American Library, New York.
2. Beauchamp, Tom L. and Childress, James F.: 1994, *Principles for Biomedical Ethics*, 4th ed., Oxford University Press, New York.
3. Brueggemann, Walter: 1989, *Finally Comes the Poet: Daring Speech for Proclamation*, Fortress, Minneapolis.
4. Cassell, Eric: 1991, *The Nature of Suffering and the Goals of Medicine*, Oxford University Press, New York.
5. Farley, Wendy: 1990, *Tragic Vision and Divine Compassion: A Contemporary Theodicy*, Westminster/John Knox Press, Louisville.
6. Fox, Matthew: 1979, *A Spirituality Named Compassion: And the Healing of the Global Village, Humpty Dumpty, and Us*, Winston Press, Minneapolis.
7. Harrison, Beverly: 1983, *Making the Connections*, Carol S. Robb (ed.), Beacon Press, Boston.
8. Holmes, Helen Bequaert and Purdy, Laura M.: 1992, *Feminist Perspectives in Medical Ethics*, Indiana University Press, Bloomington.
9. Katoppo, Marianne: 1981, *Compassionate and Free: An Asian Woman's Theology*, Orbis, Maryknoll.
10. Kemp, John: 1982, 'Hobbes on Pity and Charity', in J.G. van der Bend (ed.), *Thomas Hobbes: His View of Man*, Editions Rodopi B.V., Amsterdam.
11. Koyama, Kosuke: 1989, 'He Had Compassion', *The Christian Century* 106 (July), 5–12.
12. Kwok, Pui Lan: 1987, 'God Weeps with our Pain', in John S. Pobee and Barbel von Wartenbuerg-Potter (eds.), *New Eyes for Reading: Biblical and Theological Reflections by Women from the Third World*, Claretian Publications, Quezon City.
13. Kyung, Chung Hyun: 1990, *Struggle to be the Sun Again*, Orbis, Maryknoll.
14. Lebacqz, Karen: 1995, 'Feminism', in Warren T. Reich (ed.), *The Encyclopedia of Bioethics*, second ed., MacMillan, New York.
15. Morton, Nelle: 1985, *The Journey is Home*, Beacon Press, Boston.
16. Noddings, Nel: 1984, *Caring: A Feminine Approach to Ethics and Moral Education*, University of California Press, Berkeley.
17. Nouwen, Henri M., McNeill, Donald P. and Morrison, Douglas A.: 1982, *Compassion: A Reflection on the Christian Life*, Doubleday, New York.
18. Pellegrino, Edmund and Thomasma, David: 1988, *For the Patient's Good: The Restoration of Beneficence in Health Care*, Oxford University Press, New York.
19. Reich, Warren Thomas: 1989, 'Speaking of Suffering: A Moral Account of Compassion', *Soundings* 67 (Spring), 1.
20. Ruddick, Sara: 1986, 'Maternal Thinking', in Marilyn Pearsall (ed.), *Women and Values: Readings in Recent Feminist Philosophy*, Wadsworth, New York.
21. Sacks, Oliver: 1984, *A Leg to Stand On*, Harper and Row, New York.
22. Scary, Elaine: 1985, *The Body in Pain*, Oxford University Press, New York.
23. Scott, Waldron: 1987, 'Mercy and Social Transformation', in Samuel Vinay and Chris Sugden (eds.), *The Church in Response to Human Need*, William B. Eerdmans, Grand Rapids.
24. Sherwin, Susan: 1992, *No Longer Patient: Feminist Ethics and Health Care*, Temple University Press, Philadelphia.
25. Sister Mary Celeste: 1963, 'The Virtue of Mercy', *Review for Religious* 22 (May), 3.
26. Soelle, Dorothee: 1975, *Suffering*, Robert and Rita Kimber (trans.), Westminster Press, Philadelphia.
27. Song, Choan Seng: 1982, *The Compassionate God*, Orbis, Maryknoll.
28. Trible, Phyllis: 1978, *God and the Rhetoric of Sexuality*, Fortress, Philadelphia.
29. Underwood, Ralph L.: 1985, *Empathy and Confrontation in Pastoral Care*, Fortess, Philadelphia.
30. *Webster's New Collegiate Dictionary*: 1974, G. and C. Merriam, Co., Springfield, MA.

SECTION II

Practices, Concepts, Methods, and Theories

TIMOTHY E. MADISON

Tunnel Vision or Moral Discourse?
An Insider's View of Bioethics in a Medical Center

THE HISTORY OF ETHICS

While preparing an introduction to bioethics for a group of clinicians, I discovered an intriguing document in the files of my hospital's bioethics committee. It was an unpublished, unsigned article entitled, "The History of Ethics". Amazingly enough this historical survey was contained in one page! A scan of its contents provided a distressing explanation for its brevity. The "history of ethics" had been reduced to a series of U.S. court cases and medical dilemmas beginning shortly after World War II. The cases were described in the language of medical analysis and legal outcome.

THE "HEALTH CARE REFORM" OF BIOETHICS

Three years of experience as co-chair of a hospital bioethics committee has convinced me that this sort of tunnel vision is prevalent in the dialogue and literature often utilized in the clinical setting. Bioethics has experienced its own "health care reform". Careful analysis and character formation, which are considered expensive and time-consuming, are trimmed away as excessive fat. In place of formative methods whereby teachers equip students to become teachers, there is the problem-solving consultant who can read medical records, interview patients, families, and health care staff, and offer nonbinding recommendations. Some of the most widely read bioethics literature now cater to the needs of the bioethics consultant and his or her busy clinician clients. Examples of periodicals include *Medical Ethics Advisor* by American Health Consultants and *Hospital Ethics* by the American Hospital Association.

Two recent books illustrating this accommodation are *Ethics Consultation: A Practical Guide* [12] and *Values in Conflict: Resolving Ethical Issues in Health Care* [1]. The former is a manual for bioethical consultants written by two physicians. Carefully detailed instructions are provided on topics ranging from the proper protocol for clinical consultations to methods of

99

E. E. Shelp (ed.), Secular Bioethics in Theological Perspective, 99–114.
© 1996 Kluwer Academic Publishers. Printed in the Netherlands.

reimbursement for bioethics consultations. Training for such work should consist, at best, of a medical degree and further specialization in bioethics.

The latter focuses on the legal translations of certain ethical principles. Sections are devoted to topics such as informed consent, advance directives, end-of-life policies, confidentiality, and structural analysis of the health care system. The expected audience consists of busy administrators thus all the above topics and more are covered in seventy pages. What is necessarily omitted from such a work is any sense of the rich history and complexities of ethical reflection in the health care context. It bears a troubling similarity to the aforementioned "History of Ethics". The reader is left with the impression, for example, that when a medical center constructs a legal DNR policy then it has met most, if not all, of its moral obligations. Anyone who has ever stood beside a patient, family, and staff making end-of-life decisions will testify that institutional policies may sometimes alleviate anxiety but they seldom fulfill the moral agents' responsibilities to each other or themselves.

Both these books are what they claim to be. They are not misleading. The danger they represent is the effort to streamline ethical analysis to a method of problem-solving suited to a secular, technical context. Such problem-solving too often takes the form of a negotiated agreement between competing legal entities based on a consultant's expert testimony rather than a nuanced moral response based on self-critical ethical analysis.

The proliferation of advance directives is another example. In this case a legal document entitled "Directive to Physician" is introduced into the physician-patient relationship in hopes of progress toward the resolution of end-of-life dilemmas. The bioethical consultant may be brought in to read the document and deem if it accurately reflects the preferences of an autonomous agent. If so the physician is legally bound to comply with the patient's stated wishes or transfer the patient to a physician who will. What is missing is the necessary messiness of dialogue between the involved parties in which moral values are presented and defended in an open fashion. Missing is an empathetic understanding of how these values fit, or do not fit, the patient's personal narrative. Add to this missing list the analysis of how the health care staff's moral assumptions impact their prognosis and treatment plan. To arrive at such valuable insights, careful reflection and education will have to take place in a structured manner. Such a process will take time. It will require a broader community of inquiry than a few bioethical technicians. The outcomes will be less controllable. Movement, however, toward better equipped moral agents may become possible. That movement seems unlikely amid an approach to bioethics which highlights paid specialists and institutional policies.

A COMMUNITY OF MORAL DISCOURSE

Theological ethics offers a highly promising concept to bioethicists who seek such movement in a secular context. This gift is James Gustafson's community of moral discourse (hereafter abbreviated as CMD to accommodate busy readers). Gustafson is "loathe to contribute to the redundance in the ethical literature" ([9], p. 183); and I apologize to him for reiterating some points he has made far better. I am convinced, however, that the prevalent applications of ethics to health care are omitting a valuable resource in Gustafson's work perhaps due to its lengthy analysis and/or its theological themes.

No attempt will be made to summarize Gustafson's contributions to bioethics much less to ethics. Some very fine expositions on this topic may be found elsewhere [2], [16]. Certainly Gustafson is not the first major thinker to address bioethical issues and is quick to credit predecessors such as Aquinas, Barth, and H. Richard Niebuhr as well as contemporaries such as Paul Ramsey. Gustafson, however, is among the first theological ethicists to examine some of the difficult questions of health care and research.

In his writings, Gustafson makes explicit certain ground-of-meaning assumptions which lead to his advocacy of CMDs in bioethics. First, he asserts:

The purpose of medicine is to sustain human physical and mental well-being. . . . health is not an end in itself; it is a necessary condition for human functioning to realize other purposes, and for the capacity to exercise human agency ([8], p. 253).

Medicine is a culture's instrument for defending its members against agency-debilitating disease and death. The key is to recognize that health care's moral end is not the continuation of physical existence for its own sake but for the continuation of responsible agency.

Second, moral agency is a complex, multi-layered activity. It includes:

basic postures toward the world of which we are a part, our dispositions toward the others with whom we interact, as well as our reflection on what we should do and our actions that seek to fulfill purposes ([8], p. 3).

As such, moral activity is inherently interactional. It is within relationships that persons experience moral obligation. The various possibilities faced by moral agents are no less dynamic than the specifics of human relationships. These possibilities are limited by the agent's characteristics and specific context. These limits are themselves fertile ground for reflection and action ([16], p. 107).

The "Why be moral?" question is answered by Gustafson in theological terms. Human moral activity is obliged and enabled by God's activity in "intending the well-being of creation" ([16], p. 112). Herein lies the crucial difference between Gustafson's theocentric ethics and most other ethical systems. All ethical pursuits – including health care – are judged by the standard of service to God, not human well-being, happiness, or excellence. Gustafson puts it bluntly, "Man – individual human beings, human communities, and the human species – is not the ultimate center of value" ([8], p. 4). This dramatic shift in perspective, if accepted, requires more than a set of religious beliefs and loyalties. Theocentric ethics necessitates deeply integrated religious dispositions and affections from which ethics emerges.

Third, these dispositions and affections seldom develop in isolation but within communities of moral formation. Moral reflection and activity, therefore, require a community. This is a consistent theme in Gustafson's writings dating back to his first book, *Treasure in Earthen Vessels*. In his most recent work he writes: "The maintenance of a community and its symbols is a necessary condition for the preservation and even reconstruction of the piety and theology I have developed" ([7], p. 324). Neither moral analysis nor consultation is the practice of an isolated ideal observer. Even individuals who make significant moral contributions emerge from communities which deeply influence the types of moral agents they become.

There are many communities which consciously or unconsciously engage in moral formation. Gustafson asserts, however, that moral participation in the complex issues facing humanity seldom occur accidentally. Participation more often emerges from a community intentionally and consistently committed to ethical discourse and action. This is what Gustafson calls a CMD. It is defined as

> a gathering of people with the explicit intention to survey and critically discuss their personal and social responsibility in the light of moral convictions about which there is some consensus and to which there is some loyalty ([5], p. 84).

Notice Gustafson's stress on the intentionality required for moral discourse. No gathering – whether a hospital bioethics committee or a presidential commission – is automatically a CMD. Moral discourse requires focused interaction on the moral dimensions of complex situations. As such it requires several elements: a community able to articulate a common loyalty; people willing to engage in dialogue on basic understandings of what is good and right; the language of ethics; and the willingness to risk change amid unpredictable interactions.

Such a community can lead to constructive collective action but more importantly it can result in moral agents who perceive their context insightfully and humanely. Their moral assumptions can be revealed and critically discussed. They can dialogue with different perspectives, learning both empathy and self-determination. They can live in the world – even a hospital – with a broadened, deepened understanding of life's moral dimensions which better equips them to face life's moral dilemmas.

A CMD requires a difficult reorientation and a transformation of the very phrasing of moral questions. Individuals and communities cannot act solely from self-interest. Instead they first must ask what knowledge of the larger whole is available. Then they must investigate what can be known about God's purposes for that greater whole of which we are a part ([15], pp. 40–41).

Try fitting that language on the agenda of a hospital bioethics committee. More importantly, try communicating this perspective and these concepts to the clinicians, administrators, attorneys, legislators, patients and families who must resolve bioethical issues in intensely stressful circumstances.

One task which secular bioethics consistently has performed better than theological bioethics is the presentation of concepts in a fashion that busy, distressed moral agents can manage. The identification of certain principles and rules as being primary has put these difficult concepts on the lips of physicians, nurses, and administrators. Diagrams of decision-making schemes can be found throughout health care literature.

Can theological ethical concepts – like the CMD – move into the language and comprehension of medical centers and court rooms where bioethical dilemmas are resolved? What follows is movement toward one such attempt. It is a risky venture because I may end up with a description of a CMD which ethicists defame as over-simplified and clinicians reject as obscure. The risk is worthwhile, however, because secular ethics with all its strengths in consultations and seminars has not resolved the ethical dilemmas facing health care practice, research, and distribution. The CMD is a theological concept which, at least, can complement the progress secular bioethics has made in the clinical arena.

Evidence of a need for CMDs is found in the "gee whiz" approach to health care dilemmas which dominates not only the mass media but also the literature and training of hospital ethics committees. William May echoes this concern in his discussion of "quandary ethics" ([13], pp. 13 ff.). This approach is characterized by the presentation of a difficult case with competing values and personalities which impress people with the case's complexities but does not equip them to respond ethically. The cases move to court where legal questions may be resolved but not the moral questions. Too often the public responds to such cases with statements such as, "I guess we will have to let each individual's sense of right and wrong decide the moral questions". In bioethics this assertion has been translated into the legal language of autonomy. Taken seriously, however, autonomy as panacea leads to an anarchy of unreflective morality. Simply defending a person's autonomy does not challenge the moral participant to analyze critically the origin and validity of his or her assumptions. Ethics surrenders to cliches and illusions of ideal moral advisors. The dangers include the inability to confront complex bio-ethical issues in areas such as the Human Genome Project ([9], p. 190).

There is a better way. The CMD is a place where complex issues can be examined and anticipated, and where the people who must face them can be equipped for constructive participation. Why would a CMD help? Gustafson offers five replies: (1) Choices are more likely – though not inevitably – to value correctly between good and evil if reflection occurs with other participants. (2) No individual has the exhaustive knowledge required to be absolutely morally self-reliant. (3) Personal biases must be challenged by a wide variety of alternatives. (4) Mutual understanding by people holding divergent values is possible when they approach a consensus process with the willingness to change their minds. (5) Such dialogue can and should involve both theorists and clinicians ([8], pp. 316–317).

In theological terms, the CMD is a necessary corrective to human sinfulness as seen in our denial of finitude and interdependence. In sociological terms, the CMD affirms the role of communities in the formation of moral values and the participation of its members in the dialogue and activity of ethics. Morality has as its focus the interactions of relationships. Bioethics is no different. The CMD offers a structured forum where those interactions may occur and be examined in movement toward an appropriate response.

Why employ CMDs in health care? Gustafson replies:

The enterprise of medicine, while frequently practiced by isolated individuals, is a communal one, and the formation of attitudes takes place within professional communities ([6], p. 75).

The practice of medicine emerges from a formative community. Why should ethical reflection on health care occur any differently?

Gustafson realizes that clinicians are not going to flock to abstract discussions about justice ([3], p. 98). Nonetheless he asserts that CMDs are necessary for a careful examination of the nature of human moral activity. CMDs, for example, offer to the discussion of genetics:

some of the specialization and diversity needed to include the many considerations that should be addressed. Media presentations often focus on the more dramatic and esoteric possibilities . . . There is no substitute, however, for the hard work of cross-disciplinary discourse which involves sufficient knowledge of the information, concepts, and ways of thinking represented by specialties and interests ([9], pp. 199–200).

FOUR DIMENSIONS: DATA COLLECTION

Gustafson does not prescribe a process for a CMD to follow in its work, perhaps because he wants to avoid advocating some cookbook to successful moral decision-making. Ethics is not that simple. I propose, however, that four dimensions of the work of a CMD can be gleaned from his writings. They are presented as dimensions because, while bearing distinct qualities, they often overlap in the real work of the CMD. They are not chronological stages. These dimensions are the collection of data, open dialogue, continual reflection, and participation.

The collection of data is a crucial foundation for the CMD process for without adequate attention to gathering relevant information, moral responses reflect biases more than understanding. This dimension necessitates careful research and subsequent dialogue within and outside the CMD regarding the pertinence of the data for the context under inquiry. Such dialogue requires one to invite people with diverse specialties and interests to the table. An open flow of information should be established.

Bioethical inquiry, therefore, begins with a collection of the data surrounding a particular context. Such data includes, but is not limited to, medical diagnosis and prognosis. Too often such medical data is accepted at face value. This is particularly common when non-physicians are intimidated by a deluge of technical terms. Gustafson calls data collectors to dig a bit deeper into the presentation. He advocates three questions for data: (1) What information is relevant to the moral context? (2) What interpretation within a scientific field is assumed and why? (3) What are the value biases behind the empirical presentation ([11], p. 226)? The value of adopting such analytical questions is that they help participants avoid the extremes either of ignoring empirical data because it is too technical or discarding ethical analysis of that data due to the presenter's authority claims.

On the one hand, devoting time to data collection helps moral participants to avoid simplistic applications of some singular, universal abstraction to a particular context. A community's moral interactions are complex affairs which require a careful presentation of the perceptions and situational dynamics which impact each case's participants.

On the other hand, the CMD needs to be aware of certain data collection hazards. First, presentations of data are to be read with a critical eye. Beware of sources which claim authority by just reporting the "facts". Information about moral events is still human information. Second, data collection is only an initial step in moral discourse. Investigative reporting does not replace intentional analysis. Third, the CMD must avoid becoming bogged down in exhaustive research. As finite beings there is a limit to how much a community can accumulate and comprehend. Sometimes the moral context will necessitate an expedient response. In any case, attempts to know everything before acting simply become avoidance of moral participation.

FOUR DIMENSIONS: OPEN DIALOGUE

The CMD's second dimension is open dialogue. This dimension is warranted by the absence of a static, harmonious moral order which would lead to consentient moral judgments. The "open" nature of this dialogue is described by a Gustafson critic as follows:

> One assumes some part of one's web [of insight] is true, even as one subjects another to criticism. In principle, no beliefs are immune to revision; one aims at coherence but not in the sense of mere consistency, for one tests, changes, adjusts in an ongoing process of explanation and justification ([14], p. 123).

Since no individual or community can claim a corner on moral truth, serious research and discourse are prerequisite to the formation of well-developed ethical arguments.

Open dialogue in bioethics includes appeals not only to medical data but also to religious beliefs, subjective feelings, and minority viewpoints. It is crucial to understand, however, that a gathering of people who address any or all of these subjects is not automatically a CMD. Moral discourse requires the use of the language of ethics. Ethical terminology must be employed and each term should be analyzed in terms of its benefits and limits ([3], p. 96). The language of ethics is evident, for example, when a bioethics committee moves beyond discussing an AIDS patient's diagnosis to raise questions like the following: What obligation does the physician have to the person with AIDS? What is the fairest means of distributing scarce medical resources among patients in an era of AIDS? What are the patient's moral values and how might the treatment plan respect them? How do the moral values of the health care team members influence this patient's treatment?

Maintaining open dialogue also means not restricting membership to a panel of experts who resolve presented problems for the larger community. The purpose of a CMD is to enhance the capacity for moral participation among those who bear responsibility for those decisions:

> It is to help form the "consciences" of persons, to educate their rational activity, to enable them to think more clearly and thoroughly about the moral dimensions of aspects of life in the world. It is to hone more sharply their moral thinking from which choices and actions in part flow ([8], p. 317).

Open dialogue, therefore, facilitates individual accountability. Such dialogue will seem risky and bothersome to some health care professionals. They are accustomed to making rapid judgments and are pressured to do so by patients, families, administrators, and insurers. In the hierarchy of health care decision-making, open dialogue may border on insubordination. The benefit, however, is that open dialogue prevents certain powerful individuals and moral assumptions from being omitted from critical analysis.

Open dialogue appeals to the courage of health care professionals who are accustomed to taking risks in order to benefit their patients. Moral discourse rewards such courage with improved capacities for moral understandings and increased opportunities for creative collective action.

When applied to bioethics, open dialogue becomes helpful precisely at the point clinicians feel trapped in a dilemma: interpersonal conflict. In routine medical situations, general moral rules apply which require little or no discussion. A patient under anesthesia, for example, is not to be abused or exploited. Violations of such moral rules have punitive legal consequences due to the broad social consensus which supports them. A process of open dialogue is unnecessary in such cases. The focus of contemporary bioethics, however, has come to rest on more ambiguous circumstances. These cases are characterized typically by conflicts both of personalities and moral values. Ethical analysis then becomes less a matter of listing moral rules and more a practice of balancing competing moral assertions and loyalties. The benefit of open dialogue is its capacity to lift out "the conflicts of values and principles that are implied in human experience, and seek to revise and correct their ordering so that actions can be directed more properly" ([8], p. 301).

For example, what if a variety of CMDs approached the highly polarized issue of abortion, not as an exercise in choosing sides but as an exercise in open dialogue? After careful research, an exploration of the value of human life would ensue. Gustafson would argue that this value lies in the indispensability of physical existence for all activities which humans value. The burden of proof, therefore, is always on anyone who would end a human life ([4], p. 140). Open dialogue could focus on the presence or absence of exceptionally good reasons for ending a life and what, if any, distinctions between prenatal and postnatal life might be justified. The full range of moral experience and interests would be brought to bear amid weighty discourse. Open dialogue might replace a few of the angry accusations.

A related example involves the balancing of individual rights versus the well-being of the common good. Contemporary bioethics bears a strong inclination toward the preeminence of individual autonomy. There are some very good reasons for that inclination, including the patient's vulnerability in the clinical context and the continuing revelations of researcher's abuses of human subjects. At the same time some recent end-of-life cases have demonstrated that simply upholding patient autonomy does not eliminate the need for ethical discourse. For example, cases in which families insist on continuing aggressive interventions for patients beyond the benefit of such care create difficult dilemmas for clinicians. Not only do physicians and hospitals desire to respond positively to family requests but they also wish to avoid punitive litigation. A balancing act ensues in which autonomy, beneficence, distributive justice, etc. are weighed against each other in hopes of reaching a satisfactory resolution. Unfortunately much of this dialogue occurs in conflict-driven isolation rather than in a community where open discourse between all parties is affirmed.

When open dialogue occurs, one discovers that:

> there is never only one outcome of a human action; there may be one immediate recipient but in the subsequent interactions there are many recipients in the larger patterns and processes of interdependence ([8], p. 282).

Individual autonomy is limited by interdependence. The respect for an individual's right to choose is balanced by respect for the community of which that individual is a part as well as other communities which may bear the weight of consequences. Our experience in families should be ample to demonstrate the liabilities of an approach to bioethics which makes all moral values subservient to autonomy ([16], p. 119).

Such weighing of values and principles requires careful argumentation, but open dialogue is misunderstood if one reduces it to a purely rational procedure. Gustafson advocates discernment rather than problem-solving. Much of the focus in the orientation of clinicians to bioethics has to do with the cognitive steps by which people make choices. The objective is to move individual patients or proxies toward the capacity to make autonomous rational decisions based on their own moral values. If the person is conscious and competent then bioethics is a process of clarifying information and raising options. The agent makes his or her choices and, whenever possible, the health care team accommodates those decisions.

As far back as 1971, Gustafson criticized this form of ethical analysis as incomplete. To treat ethics as solely a matter of isolated autonomous choices is to assume:

> a mode of life which is largely one of problem-solving, of achievement of specific purposes or ends, and tends to slip into a flat, mechanistic view of experience. It reduces the sense of awe and wonder ([4], p. 147).

Ethical analysis is instead a process of discernment, of perceiving a fitting way to understand and participate in a specific context.

Discernment is *not* any of the following: (1) a mechanical itemization of cases into general categories according to brief observations; (2) the establishment of a set of principles into which each case is forced without attention to the complexity of detail; (3) exhaustive research of detail without attention to relevance; (4) deep feelings expressed uncritically; or (5) stubborn allegiance to a group of rules or authority figures ([10], pp. 20–23).

Discernment is the ability to interpret an event, idea, or policy with an insightful and imaginative grasp of the interrelationships between the involved parties. It involves a careful description of the perceived data, enough self-understanding to reveal how one's own assumptions and loyalties impact that perception, and a reasonable argument to support that perception. Discernment, therefore, is both a rational and affective process. The discerner possesses both a clear head and an adept intuition. He or she develops the ability to articulate a broad range of alternatives and potential consequences, being careful to examine the subtle nuances hidden in moral language.

Discernment challenges both empirical problem-solvers, who trust primarily their scientific observations, and moralists, who trust primarily their abstract principles or loyalties, to appreciate a broader range of resources for ethical analysis. Gustafson appeals to Barth's "points to be considered" in order to affirm the role of reasoned reflection on a variety of subjects ([8], p. 303). These points are examined in terms of their relevance and usefulness to the issue at hand. Emotions, experiences, and religious faith along with reason weigh heavily in this process, but discernment is the whole of which these elements are a part.

The theological question which drives discernment is "What is God enabling and requiring us to do?" Discernment includes interpretive skills, personal attributes and dispositions, and appeals to basic moral principles but the process also assumes the presence of a divine Other who both obliges humans to reflect ethically on their context and enables them to participate. Allen Verhey sums it up well: every other piece of discernment is qualified by "the experience of the presence and power of God and by reflection concerning God's character, purposes, and relations with the world" ([16], p. 109).

Religious assumptions are a part of discernment. Making that fact clear-cut will render this form of dialogue difficult for those who hold a taboo against religious language in moral discourse. On the other hand, those religious assumptions are subject to analysis and revision which may prove discomforting for those who assume that religious loyalties alone enable them to resolve moral dilemmas. Religious language is a frequent companion to bioethical conversations in the clinical setting. For those seeking to integrate that language and their own theological reflections into moral discourse without being limited to a particular group's rules, discernment is an excellent tool.

Prevalent amid clinical conversations regarding bioethics is the attempt to utilize a list of principles into which a context can be squeezed to reduce moral ambiguity. The irony is that such a process is contrary to the art of a perceptive medical clinician who judges the significance not only of the pieces of diagnostic data but also the whole being and the interrelationship of the parts ([7], pp. 329–330), ([3], p. 92). Moral discernment is an art of perception and reflection, understanding the dynamics of the parts and the whole. Open dialogue is the arena in which discernment can develop. Discernment leads to a moment of judgment which is an informed intuition rather than a lock-step conclusion which eliminates ambiguity. It is not a panacea for dilemmas but a life-long search for a moral life. What is eliminated is any illusion that the basic conflicts which have created the bioethics industry can be resolved with finality. That would require the eyes of God.

Four Dimensions: Continual Reflection

A crucial element of open dialogue is that despite a community's devotion to research, character formation, and discernment, the risk and tragedy which drive moral discourse cannot be eliminated. Gustafson makes this point in a most straightforward manner:

> Known benefits must be risked for unknown; known harms can sometimes be avoided only by risking known benefits. Action is by finite human agents; it is inherently risky and frequently unavoidably tragic in the strongest sense ([6], pp. 46–47).

Moral analysis must include an ongoing examination of the suffering and destruction which results from attempts at ethical intervention. Continual reflection is this third dimension of a CMD.

It is necessary that any group continually reflect on their discourse and activity because no matter how well-intentioned the members of that group may be, they remain limited human beings who cannot grasp the good with absolute certainty. Grave reflection and vigorous justification are required to balance out the impact of self-interest and group loyalties. Continual reflection is a self-critical process which includes mercy, self-restraint, and, at times, self-sacrifice.

For the group engaged in bioethics, continual reflection compels questions about the process being employed. Has the data collection emerged from two or three resources which share a particular moral position? Is the group dialogue being dominated by an individual or by a particular profession? Is the group's activity open to a variety of perceptions or does it lean too much on one expert to shape its participation? What were the consequences of the group's previous interventions (whether by direct action, consultation, policy, or education)? The analysis of consequences should include individual patients and health care staff as well as the institution, health care systems, larger community, and the environment.

Continual reflection does not mean that the CMD paralyzes itself with obsessions about whether or not it is doing the right thing. Such self-preoccupation cripples the capacity for participation. Continual reflection means that one's research, dialogue, and action are tempered by an ongoing awareness that such efforts remain human efforts. Thus its arguments and conclusions need to be balanced by a consistent process of self-critical reflection. Continual reflection is a self-critical filter through which a group's moral perceptions and participation pass. That filter is continually in place or else the group endangers its identity as a forum wherein open dialogue and innovative moral activity are possible.

In this era of centralized institutional power structures in the health care industry, continual reflection is as necessary as it is difficult. The bioethics group within such organizations should count among its highest priorities the analysis of loyalties, process, power distribution, and consequences. The moral character to be shaped in the CMD process, therefore, will include courage.

FOUR DIMENSIONS: PARTICIPATION

The fourth dimension is identified by a term used throughout this discussion: participation. Bioethicists are not to be content with uttering prophetic pronouncements or commanding unswerving allegiance to the preservation of moral absolutes. The person talking about bioethics should be engaged in the context being discussed.

The concept of participation evolved in Gustafson's thought beginning with a theological argument against human beings as "co-creators". His reformed tradition background asserts the sinful nature of human activity and perception. To claim a creative capacity for human beings on par with the creative, ordering activity of God is to claim far too much for finite beings. Building on H.R. Niebuhr's concept of ethics as response, Gustafson initially argued that humans could become "co-actors" through responsible activity amid divinely created possibilities ([6], p. 52).

In his recent work, Gustafson has refined co-action to participation. The use of this term emphasizes that humans are not mere spectators of the moral context. Neither are they simply waiting to react to some perceived context external to their own lives. Participants begin with the assumption that they are interacting with other parties and structures, each of which is a part of some whole. Although humans cannot claim a clear perception of "The whole, that is, the totality of all things", they can cooperatively discern the broader spheres in which ethical issues emerge ([8], p. 15). Alluding again to theology, Gustafson compares participation to stewardship in which humans are "temporary, responsible custodians of, and contributors to, the realms in which [they] participate" ([8], p. 145). Participation affirms the human capacity for intentional, innovative involvement in a moral context while denying any claim to a perfect, once-and-for-all resolution to that context.

The dimension of participation highlights the balance to be maintained in any ethical endeavor. There is no ideal abstract theory or principle from which clinicians can deduct resolutions to dilemmas in treatment withdrawals or health care reform. On the other hand, bioethics cannot be reduced to some internal, immediate emotive reaction which dictates right and good in each situation. Participation requires feelings to be balanced by rational arguments, and both to be tempered by shared experience and community traditions of which the specific context is a part.

At the heart of participation is the balance to be maintained between human limitations and possibilities. Gustafson sums up this crucial point:

> The concept of man as participant not only discloses our limits; it also affirms our powers and capacities to intervene in the patterns and processes in accordance with our intention to achieve beneficial ends ([8], p. 280).

Claims of moral certainty are to be challenged but so too are excuses of moral ignorance. At some point, the community must move from research, discerning dialogue, and reflection to take action with the intention of changing the context for the better.

Participation as a defining dimension is a helpful balance to the current emphasis on consultation in bioethics. Gustafson clarifies the role of the ethicist in the clinical setting as follows:

> The practical function of the ethician is not primarily to prescribe and proscribe the conduct of others, but to enable them to make informed choices. Agents who have responsibility for particular spheres of interdependence and action must be accountable for the choices they make. . . . The function of the ethician is to broaden and deepen the capacities of others to make morally responsible choices ([8], p. 315).

There is an assigned priority given to the character formation of the clinicians engaged in bioethical discernment. The development of certain dispositions among clinicians is as important – if not more important – than the resolution of particular difficult cases.

Such an emphasis differs from current instruction in bioethical consultation. An example is provided in a recent work by two physicians who outline the process of bioethical consultation as follows: (1) interview clinicians and identify the issues; (2) interview the nurses and family, and read the medical record; (3) interview and examine the patient; (4) conduct family meetings; (5) prepare an ethical analysis; (6) report to the requesting physician and make four recommendations; and (7) follow-up with the patient ([12], pp. 4 ff.).

The focus in this explanation of consultation is on resolving a problem identified by a physician. The consultant is a bivocational expert in medicine and bioethics who can glean from medical data and lay interviews the pertinent ethical issues. He or she can then apply ethical principles or rules and make recommendations. The physician has the responsibility for making choices.

The participation dimension brings into question some of the assumptions of this model. Claims of ethical expertise are tempered by an awareness of the broad scope of ethics and every individual's limitations in perception and interpretation. The attempt to approach bioethics as one more medical specialty or graduate degree discounts the all-embracing nature of ethical inquiry. It slights the responsibility to participate, not only as one who sifts through an advisor's recommendations but also as an actor in the process of dialogue and self-critical reflection. Delegating the discernment process to an identified bioethical specialist may save time but it risks abandoning any opportunity for the formation of a genuine CMD. It may create a career opportunity for those interested in ethics and medicine while ignoring the responsibility not only to develop their own moral sensitivities but also those capacities among their colleagues in the clinical setting.

The courage to reflect critically on the clinician's own involvement in the larger wholes of which a case is a part will not likely result from problem-solving interviews with a consultant. The possibility for character formation along with ethical interventions is more likely to occur within a group which intentionally makes those activities their agenda. Both consultation and participation leave the final responsibility for choice and action with the context's principal agents. A significant difference, however, exists in the application of this point. Consultation, especially as a financial venture, fosters dependence upon a bioethics expert. Participation, both during and before dilemmas emerge, takes consistent initiative to equip human beings for moral responsibility in the health care context.

CONCLUSION

Gustafson's concept of a CMD can and should be appropriated for bioethical inquiry in the clinical setting. Such an effort may include bioethics committees and consultants but it will require a different level of intentionality and activity than is found in many medical centers. Specifically, a CMD will involve four dimensions. Data collection will be an ongoing process as community members research not only particular cases but the larger wholes of which those cases are a part. They will read books and journals so as at least to be better informed in the field of bioethics than those who come to them for assistance. Open dialogue will ensue characterized by careful reasoning, an affirmation of affective contributions, and the intentional use of ethical language. Continual reflection will temper all discussion and action as group members self-critically analyze the loyalties and limits they bring to the inquiry. Participation will result both in the traditional forms of case consultation, policy writing, and educational programs but also in the character formation of community members and colleagues.

These dimensions with their theological underpinnings can make a significant contribution to secular expressions of bioethics. Bioethics is not a technique to be mastered by a few experts who study medicine and recent court cases and who then recommend what is right, just, and good to those in authority. Bioethics is one piece of a lengthy human inquiry into what is right, just, and good which has been undertaken by members of every community utilizing their reason, experience, feelings, and faith. Participation is the joining of that historical inquiry. Participants cannot help but be changed by the process.

Valley Baptist Medical Center
Harlingen, Texas, U.S.A.

BIBLIOGRAPHY

1. American Hospital Association: 1994, *Values in Conflict: Resolving Ethical Issues in Health Care*, Second Edition.
2. Beckley, H.R. and C.M. Swezey (eds.): 1988, *James M. Gustafson's Theocentric Ethics*, Mercer University Press, Macon, Georgia.
3. Gustafson, J.M.: 1991, 'All Things in Relation to God: An Interview with James M. Gustafson', *Second Opinion* 16 (March), 80–107.
4. Gustafson, J.M.: 1971, *Christian Ethics and the Community*, Pilgrim Press, Philadelphia.
5. Gustafson, J.M.: 1970, *The Church as Moral Decision-Maker*, Pilgrim Press, Philadelphia.
6. Gustafson, J.M.: 1975, *The Contributions of Theology to Medical Ethics*, Marquette University Press, Milwaukee.
7. Gustafson, J.M.: 1981, *Ethics from a Theocentric Perspective*, Vol. I, University of Chicago Press, Chicago.
8. Gustafson, J.M.: 1984, *Ethics from a Theocentric Perspective*, Vol. II, University of Chicago Press, Chicago.
9. Gustafson, J.M.: 1992, 'Genetic Therapy: Ethical and Religious Reflections', *Journal of Contemporary Health, Law and Policy* 8, 183–200.
10. Gustafson, J.M.: 1968, 'Moral Discernment in the Christian Life', in G.H. Outka and P. Ramsey (eds.), *Norm and Context in Christian Ethics*, Charles Scribner's Sons, New York, pp. 17–36.

11. Gustafson, J.M.: 1974, *Theology and Christian Ethics*, Pilgrim Press, Philadelphia.
12. La Puma, J. and D. Schiedermayer: 1994, *Ethics Consultation: A Practical Guide*, Jones and Bartlett, Boston.
13. May, W.F.: 1991, *The Patient's Ordeal*, Indiana University Press, Indianapolis.
14. Reeder, J.P., Jr.: 1988, 'The Dependence of Ethics', in H.R. Beckley and C.M. Swezey (eds.), *James M. Gustafson's Theocentric Ethics*, Mercer University Press, Macon, Georgia.
15. Sande, J.R.: 1991, 'Open Letter from a Medical Student', *Second Opinion* 16, 32–49.
16. Verhey, A.D.: 1988, 'On James M. Gustafson: Can Medical Ethics Be Christian?', *Second Opinion* 7, 104–127.

STEPHEN E. LAMMERS

The Medical Futility Discussion:
Some Theological Suggestions

INTRODUCTION

In recent years, medical futility has become one of the most discussed issues in secular bioethics. In the clinical setting, the interest in this topic has extended from discussion of basic matters, i.e. what does "futility" mean, to debates about what then ought to be done when a proposed treatment is futile according to the particular definition under discussion [5]. The matter has taken institutional shape in that members of hospital ethics committees often discuss whether their hospital should develop a futility policy. One can find full bibliographies on the subject, a sure sign of the maturing of the topic within the clinical setting [4].

The debate does not stop within the clinical setting. The discussion of futility engages academics with an interest in secular bioethics as well. For them, the discussion opens interesting theoretical questions as well as questions about what takes priority in medical care, physician assessments about what is appropriate to do and not to do in medicine as opposed to patient's wishes about what should be done and not done. Thus futility is one of those topics which has engaged writers both clinical and academic in secular bioethics, and as will become clear, has led to divisions between and among them.

At one level, all of this discussion is odd. Since the time of Hippocrates, physicians have been told that it is possible that disease has "overmastered the patient" and that one of their responsibilities was to inform the patient when that point had been reached. It was also part of the Hippocratic tradition that the physician's obligations to the patient would change in those circumstances. To use modern terms, the focus of the physician's efforts should be upon "care" and not upon "cure". How is it that we find ourselves in a place where there is, or seems to be, so much contention about the judgment that there is nothing more that medically can be done to cure this particular patient of their specific disease? Is it, as some have suggested, that this is simply one more manifestation of a culture which denies death [3], or are other matters at stake as well?

In the first part of this paper, I want to give an account of how we arrived

115

E. E. Shelp (ed.), Secular Bioethics in Theological Perspective, 115–128.
© 1996 Kluwer Academic Publishers. Printed in the Netherlands.

at a place where the discussion of futility could take on so much importance. Hopefully, that account will help the reader understand the power of the current paradigm in secular bioethics. Then I will try to show how a theological perspective on these matters opens some further questions for consideration. What I hope to show is that as careful and as candid many of the discussions of futility are, they will be able to make only limited progress on these matters until some other questions are raised, questions which are raised only with great difficulty, if at all, in secular bioethics as it is currently understood in America.

It is not my intention to raise these questions as if they were part of the "view from anywhere". Indeed, to proceed in that fashion would be to replicate some of the difficulties of the kind of bioethics which stands in need of critique. I write out of the Roman Catholic tradition. It is my hope that others might understand and perhaps affirm what I am proposing. In my view, the discussion of futility exposes one of the fault lines at the heart of secular bioethics. The point and purpose of medicine is what is under discussion. At one time, religious communities might have been found in the public forum making a difference in this discussion. In our current situation, I am not confident that they will be found there. In part, this is because it is not always clear that these religious communities understand what is at stake in the discussions. Even if they were to join the discussions, it is also not clear whether they would make any difference to the conversation. Before I develop these points, it is important to understand how we arrived at where we are. To that task I now turn.

How We Got Here

Futility was first discussed in the context of cardiopulmonary resuscitation (CPR) and the debates over do-not-resuscitate (DNR) orders which engaged and continue to engage clinical medicine. For the reader unfamiliar with medical practice, CPR offers a fascinating glimpse into the world of medicine. Cardiopulmonary resuscitation is a dramatic intervention, and can be initiated without a physician's order. It spread from its initial limited usage in the operating room to all areas of medicine, including emergency medicine and beyond. CPR training is now standard for many occupations. Lifeguards, for example, routinely receive training in CPR.

As dramatic and effective as CPR was, it turned out that this intervention was not effective for certain classes of patients. Within hospitals, researchers began to ask the following question: "Do people who survive this intervention live to be discharged alive from the hospital?" If they do not, even if they survive the resuscitation, the intervention is futile [12].

Medicine was thus prepared for a discussion of futility with the spread of the DNR order, which would have health care personnel forego one of medicine's most dramatic interventions. As evidence accumulated about resuscitation, physicians began to discuss with patients the poor prognosis in some cases and to seek agreement from patients for a DNR order. There was a controversy over writing DNR orders without discussions with patients and families in situations where there were few, if any, survivors of resuscitation attempts [18].

At the same time, there was a lack of clarity about the way in which the term "futility" was used. For some physicians, it meant that the intervention was physiologically futile, in that it would not achieve the announced goal of reviving the patient. For others, it meant that the intervention would have no benefit to the patient, in this case, i.e., CPR, the patient would not live to survive the hospitalization. Obviously, what benefits the patient is open to some interpretation and if the patient wanted only to live long enough to say good-bye to someone the patient had not yet seen, resuscitation may be effective, even if the patient does not survive hospitalization. Further, it was also found that physicians used the term not only to mean what was mentioned above, but to include interventions that worked only a very small number of times. Thus discussion of futility around DNR orders disclosed that physicians were not consistent about the way in which they used terms. But this is only part of the background of the discussion of futility in secular bioethics.

One of the major themes, if not the major theme, in modern medical ethics has been the insistence upon respect for the patient. Although only one of four "principles" of bioethics, respect for persons, or autonomy, rapidly gained first place [1].

There is more than one reason for this. First, modern medicine has developed some very powerful interventions which can actually cure patients from serious illness. At the same time, these interventions were thought by some to be very burdensome and especially problematic towards the end of life. Patients began to insist that they had a right to refuse medical treatments which they found burdensome and refuse them in the face of the insistence of physicians that they, the patients, accept them. This is reflected in many cases, such as the *Quinlan* case, where the family of Karen Anne Quinlan sought the permission of the court to disconnect their daughter from the ventilator to which she was attached.

Secondly, the attention to autonomy also fit what many clinicians thought was important to preserve in western medical practice. Fearful of medical care systems which they thought were not interested in what occurred with particular patients, the focus upon patient autonomy allowed these clinicians to continue to claim that the patient in front of them was the important person, not any hypothetical patients or the medical care system in general. Physicians could hold on to a reformulated sense of themselves as fiduciaries for their patients. To put this another way, the focus upon respect for persons was attractive to clinicians because it assisted them in resisting becoming crass utilitarians in their treatment of particular patients. What went unnoticed is how respect for persons fit in well with other parts of modern medical culture in capitalist societies where patients were beginning to be thought of as consumers. As we shall see, in that culture, it becomes more difficult for clinicians not to think of themselves as business people who give consumers what they want.

This discussion of the autonomy of the patient took medical ethics through its first stages of development, but along the way, a number of astute commentators noticed that some things were missing. Focusing on autonomy placed attention on *who* was to decide; it had little to say about the content of the decision [2]. Thus, among other things, what was missing was an explicit discussion of the rightness or justice of some of the claims which might be made by patients [17]. The principle of justice, one of the four principles of bioethics, did not play a major role in the discussion concerning patients and their claims. That did not appear to be a problem when terminally ill patients were demanding that treatments be withdrawn. It would become a problem when patients would demand treatments which physicians thought that patients ought not to have. As a consequence of all of this, the rights of the patient often appeared to be placed above any claims which others might make. Indeed, community claims clearly took second place. But that only became obvious with a new set of cases which brought the discussion of futility to its present state in modern medical ethics [13].

The first case which changed the focus of discussions was one which took place in Minnesota. The patient's name was Helen Wanglie, and in that case, a woman who was maintained while she was in a persistent vegetative state, was cared for by some physicians who argued that they ought not to be required to continue the highly invasive technological treatments necessary to keep Mrs. Wanglie alive. Although the care of the woman was being delivered with the agreement of the family, the physicians thought that the particular care was not only not helpful, but that it was being inappropriately given because it was futile for this particular patient. Given her underlying condition, she was not going to return to any sort of interaction with her environment. What the physicians meant by "futile" was that the treatments would not return the patient to any cognitive functioning [6].

The family objected, saying that the treatments were not futile because they kept the patient alive. The physicians sought to have someone else named as guardian; the family won the right to continue as guardian for Mrs. Wanglie.

This case really began the current debate about futility in modern medical ethics. Notice that if you take autonomy, and the extension of autonomy to surrogates seriously, the family had a powerful argument. They provided evidence that Mrs. Wanglie would have wanted everything done to sustain her life and the family was simply executing her wishes. It was wrong of the physicians, they argued, to try to have someone else named as guardian simply because they disagreed with the family's understanding of the usefulness or uselessness of the treatments given to their mother. By analogy with previous cases, it would have been wrong to ask that physicians be named guardians for Karen Anne Quinlan just because the physicians and the family disagreed over the course of Karen Anne's treatment, or in this case, withdrawal of treatment. Note also that in this case the assertion of futility did not involve a claim about the unjust use of resources. It was a claim about what the patient wanted in the first place and what was good for this patient secondarily. It was not about resource allocation to this patient rather than to other patients, nor was it about the possible waste of resources. Issues of justice did not arise because Mrs. Wanglie belonged to an HMO which was paying for her hospitalization.

Critics have pointed out that a number of matters are left unresolved with this solution to the Wanglie case. The first involves the integrity of the medical professionals who are being asked to give care to a patient when they do not see any purpose in that care since there is no benefit to the patient as the medical professionals understand benefit. This concern presumes that medicine has certain purposes and that honoring some patients' requests would subvert the purposes of medicine. If physicians and nurses had to honor all the requests made of them, then medicine would have been transformed into something quite different during the course of the twentieth century.

The second unresolved issue is that even though Mrs. Wanglie's care was covered by her HMO, the cost of her treatment was being borne by someone and she was receiving, according to some commentators, an unjust allocation of resources in that there were others who had higher claims to those medical resources than did Mrs. Wanglie. Resources used by this patient would not be available to another patient who might benefit from them. Although the futility cases often do not focus on issues of justice in a particular case, the critics point out that matters of justice are close at hand. Indeed, some commentators see physicians playing a role as stewards of the community's resources, and in that role, responsible for the careful distribution of those resources [15]. Thus, the mirror image of the first issue concerned the limits, if any, on patient demands, not simply on an individual physician but upon the health care system. For some, this is an occasion for speaking of patients' rights, for others, equitable access for all. These issues would soon split the bioethics fraternity.

This became clear in 1994. In that year, there was a fundamental division over the case of an anencephalic child who was being treated, over the objections of the physicians, for respiratory distress. Anencephalic children rarely live for a week after birth and ordinarily they are not treated aggressively. At one point, there was a proposal that these children be put on ventilators and made candidates for organ donation. In the case under discussion, the mother of the child insisted that the child be treated whenever the child developed respiratory distress. The mother cared for the child at home and took the child to the hospital for emergency respiratory care. The child's physicians went to court in order to be relieved of the obligation of putting the child on a ventilator in the emergency setting, claiming that the treatment was futile. The court disagreed and ordered the physicians to treat the child when she presented to the emergency room. The issue of cost was not central to the case.

Robert Veatch insisted that the mother was well within her rights as a parent to insist that her child be treated. Veatch argues that judgments of futility too often mask judgments of value and that physicians have no expertise in judgments of value. In such situations, physicians should follow the judgments of their patients. Therefore, the judge should find in favor of the mother [10].

Veatch's claims are consistent with what he has held earlier about physicians. Veatch thinks that too often, judgments of physicians mask value judgments under the heading of "medical judgment". When physicians are permitted to do this, they have arrogated to themselves power which they ought not rightfully to have. Only patients' judgments of value should count in terms of decision power [22].

Veatch's testimony did not go unchallenged. John Fletcher, Director of the Center for Biomedical Ethics at the University of Virginia testified for the hospital. Fletcher is concerned that if the decision were allowed to stand, dying patients could demand any medical treatment from any physician, even in the face of physician claims that the proposed treatments offered no benefits to the patients [8].

The upshot of all of this discussion is that there are debates about the various ways in which physicians use futility, the ways in which they ought to use the term, and proposals for policies for hospitals so that institutions can make their own viewpoint clear to patients before the patient comes to the institution.

However, the reason(s) why we are in the current circumstances with respect to futility also make it clear that it will be difficult to find our way out of this situation. What is clear is that most of the participants in the discussion do not want to take from patients who are seriously ill their choices with respect to their medical treatment. This does not simply mean that the patients can refuse treatments which they find burdensome but that patients can accept treatments which are experimental in the hope of benefitting themselves or others. The debate is over the claim that patients can demand treatments which physicians think offer no benefit to the patient. This claim of "no benefit" has been expressed in a number of ways. One way of understanding it is the method of Schneiderman, who says that if a treatment has not worked in the last 100 cases, it can be considered a futile treatment [21].

The alternatives for the discussion are stark. The first alternative is to continue to press for the primacy of patient decision power in the medical situation. Physicians can discuss alternatives with patients and can make recommendations, but each of these recommendations contains some kind of value judgment, and value judgments should be left in the hands of patients. In the view of proponents of this point of view, physicians simply are not qualified to make these judgments. Physicians were competent to identify the medical facts of the situation and could, when this was known, tell a patient that a particular therapeutic intervention would not meet the patient's goals. Patient goals were primary and physicians were to serve these goals where they could [22].

One variant or another of the above proposal has been defended by some of the most prominent voices in secular bioethics. Indeed, it would appear to be consistent with much of what has occurred thus far in discussions of bioethics. One other example of a defender of a pluralism of views on matters of medical value is H. Tristram Engelhardt, Jr. Engelhardt argues on libertarian grounds. Engelhardt asks us to recognize that our situation is one where no one group has the authority to name and impose their conception of a good life. This means that each of us will have to find a physician and health care setting which fits our own conceptions of a good life. However, we should not expect that others will share our particular conceptions of a good life and we should forebear trying to write our conceptions into law or rules of medical practice. Engelhardt proposes a kind of medical practice in which there is only one principle of secular bioethics, respect for liberty [7].

This alternative would put physician integrity into question, in the view of some physicians, in that physicians would have taken from them their ability to decide whether a certain treatment fit the purposes of medical practice as physicians understood that practice. This alternative was and is unpalatable to the overwhelming majority of clinicians.

But all of that is only one possibility. The second alternative was to leave to physicians their authority to define something as futile in a particular case. The difficulty with that solution was obvious; physicians are sometimes not careful with the way in which they use the term "futile" and in the face of this physician sloppiness, patients would have judgments of value which should be theirs taken from them. Patients' rights advocates are unhappy with this solution. The proposed solution as they see it would be a step backward after a period within which patients finally have some control over their medical treatments.

A third position argues that physicians are obligated to share information but are not obligated to acquiesce in patient or family demands for useless treatment [19]. This position demands a skilled clinician who would be able to communicate that the time had come to think of comfort care while also letting families know of other alternatives, should they wish to pursue them under the guidance of some other clinician. For persons familiar with the clinical setting, this is a challenging and probably impossible task, because patients and families often perceive that kind of information as confusing, conflicting, indeed contradictory. Indeed, we already have evidence from the case of the Lakeberg twins that there always will be medical centers willing to undertake procedures which the rest of the medical community deems futile by the strictest of definitions and that there will be families which will cling to any word of promise which they hear from physicians [16].

The three alternatives are thus unsatisfactory, given the assumptions of secular bioethics. One either affirms autonomy or justice or one proceeds to some kind of *laissez faire* situation. None of the alternatives has provided the basis for a consensus.

WHERE WE MIGHT GO FROM HERE

What might a theological perspective have to contribute to this discussion? Specifically, what might the Roman Catholic tradition of discourse suggest in the present circumstances?

Here it is important to be clear about what we can expect. One way of proceeding would be to propose answers to questions posed by secular bioethics. This approach may have been possible at one time, but even what counts as a question is highly contested at the present moment. A second problem is that it lets secular bioethics set the agenda. As I hope to show, theology brings a new set of questions to the discussion, questions which reformulate the agenda.

A second option would be to argue that each community should proceed to develop its own form of medical practice. This would mean that all that is possible are alternative medicines, that each community should practice its own version of medicine, and that we can no longer be confident about any agreements in the larger culture which would sustain medicine as it has been sustained in the past. For persons who are serious about this option, there is recognition that this will be a burdensome one for Christians, because they will not participate in some of the medical care which will be offered by the liberal state. Non-participation carries with it the price of nonsupport and in the world of modern medicine, that is a very high price indeed.

It is important, in my view, seriously to consider this option. The current situation may be one in which it is no longer possible to carry on conversations with secular bioethics under the terms it wishes to impose upon the conversation. One of the reasons for this is that secular bioethics presupposes the liberal state, and the liberal state is clear about the kind of medicine it wishes to support, and that is technologically intensive medicine [9].

It is important, however, to be clear about why the option of alternative medicine might be pursued. If it is pursued, it is not because respect for liberty is the highest principle but because the current forms of medical practice, and the secular bioethics which sustains these practices, are too constraining. Those persons who explore this position from within the Christian community remind us that withdrawal from problematic circumstances is an option always to be considered. This option hardly need be considered "sectarian", unless medieval monks are also understood to be sectarian as they built an alternative to the culture in which they found themselves.

A third option would be to enter into a public conversation and to try to reconstitute a public sense of the point and purposes of medicine. The effort here would be to reclaim for medicine a sense of its enterprise, with a clearer sense of the limitations which should accompany that enterprise. As one proceeds in this direction, it should be clear that the difficult issue will continue to be the institutions which undergird secular bioethics and the medicine upon which it reflects. There are many forces already at work in the medical world which would try to undercut such a conversation. For example, discussions of futility must proceed in the face of discussions of patient rights in legislatures and in courts. This means currently that physicians are sometimes required to continue treatments they judge to be both futile and burdensome.[1]

In the remainder of this essay, I will explore the third option. This option will be pursued through a number of suggestions for discussion. This option is undertaken with the realization that those persons who do secular bioethics may not be interested in the suggestions and that it may come to pass that Christians will only be able to practice an alternative medicine.

The Roman Catholic tradition is rich enough that a number of suggestions might be made relying upon it. I want to separate these suggestions, because I think that some are more central to the tradition and to the issues than others. The other point to keep in mind is that this tradition addresses different audiences. In what follows, I will try to keep clear what audience is being addressed with the particular question.

The first suggestion, one which offers an alternative perspective on the whole discussion of futility, comes out of the Roman Catholic liturgy. The focus of this suggestion is all of us who are or will be patients and the liturgical theme is the reminder of death. What appears to be lacking in many of the discussions of futility is an explicit recognition of death, so that it cannot only be recognized but confronted. Stating that decisions surrounding death are value decisions and thus should remain the patient's is finally to say not enough from the perspective of this tradition. For example, although we may, for whatever reason, continue to deny our mortality up to the moment of our death, this denial is not the sign of a good life. It is not enough to say that the decision on this matter remains with the patient. One must take a position with respect to our mortality and whether we should be preparing for our deaths. There is no neutral place to stand on this question. Indeed, in this case, neutrality is an illusion. From the viewpoint of the Catholic liturgy, our dying is something for which we ought to be preparing and our unwillingness to make this preparation part of our lives leads to a situation such as our current one. It is not surprising that patients continue to wish to live and ask physicians to help them do this, even if it means others must suffer the loss of resources which might be given to them. The denial of death, under the guise of neutrality about what constitutes a good life, is simply an illusion which is sustained by the secular bioethics of today. As such, the denial of death has to be challenged. This claim is made for all of us. It is not a policy proposal but an outlook which has to be incorporated into our lives. It is possible for some persons to live as if they were neutral on the question of whether we should affirm or deny our mortality. From the perspective of this tradition, that would be to ask to live as if an illusion were the truth.[2]

This leads to the second point from the perspective of this tradition. Questions of justice are at stake and the way in which the four principles of bioethics are interpreted directs us away from questions of justice. The futility discussion leads to consideration about the way in which resources are used at the end of life. Given the priority of autonomy, the first question asked when a patient desires a futile treatment is whether the patient or the patient's insurance company can pay for the medical treatment. This approach completely ignores the question of whether resources are being used inappropriately in this case. What is also being ignored is the fact that technologically intensive medical treatment puts a heavy burden on our health care system and pointing to patient choice in this matter is scant comfort to those who will not receive medical treatment because of those patient choices. Let me put this another way. The entertainment industry might exist to fulfill our fantasies; medicine has other purposes and these are being ignored in this scenario.

What the futility discussion has made clear is that in the context of a society which prizes individual accomplishment and decision above everything else, even the four principles of bioethics are an unstable mixture, that in the end, autonomy will trump all else. Defenders of autonomy will have to take the effect of what they argue into account, and stop paying attention only to the intent of what they say. Not only do patients demand that they be given the right to decide whether an intervention which maintains life is "worth it" but they also assert the right to interventions which simply forestall death and insist that these be provided by physicians and nurses. It is ironic that even the four principles are not able to sustain themselves as *four* principles without some other kind of social and intellectual support. The principles themselves were and are parasitic upon older conceptions which the principles were designed to replace. It thus turns out that the four principles need to be supported by a richer conception of a good life [11]. All of us interested in bioethics, secular or otherwise, need to keep this in mind.

There is another irony precisely at this point. One of the important debates in modern medicine concerns how, if at all, physicians might act as fiduciaries for patients when they are also asked to be agents for institutions, such as HMOs [21]. One of the ways for that discussion to make some progress is to focus upon the purposes of medicine and then relate that discussion of purpose to the various tasks physicians are being asked to perform. Secular bioethics appears to undercut the real tension here and simply presume a situation of distrust between patient and physician. If we are to understand physicians only as agents for insurance companies, HMOs, or themselves and thus no longer working as fiduciaries for patients, then secular bioethics has laid out one framework for understanding their activity and responding to it. If we think of physicians as having some other roles, then we must lay the intellectual groundwork for those roles. That discussion has hardly begun. To begin it, we would have to discuss what authority we think is legitimate for physicians to have [14].

A theological perspective would help us set the framework for a discussion of the issue of authority in medicine. This is difficult in the current context, because the question is framed in terms of a zero sum game; the physician gains authority at the expense of the patient's participation in the decision about life; the physician loses authority the more the patient is given authority to decide what ought to be done. There is no sense that medicine has purposes in which both *physicians and patients* participate. If both have responsibilities and authority, then each can respect the other without either feeling or being diminished.

The underlying theological issue here is the limited nature of medicine as a human enterprise. The standard humor in our culture about the physician who plays god is well known. Now we have the physician being asked to play god and declining the role. Note that this comes at a time when much of the larger culture celebrates the accomplishments of medicine without any sense of the limitations of medicine [23]. If the physician declines the role of god, as he or she ought, then it becomes incumbent upon religious communities to support those physicians who refuse to be drawn into inappropriate patient care, inappropriate being defined as care which is not consonant with the purposes of medicine. To do this, religious communities will have to have more discussions than they have had recently about the point and purpose of medicine and what the limits they think medicine has. This is one area which needs work by medical care personnel who are also participants in religious communities.

Of course, even raising the question of the "purposes of medicine" is to put forward something suspicious. One way to raise this question of the purposes of medicine is to imagine that a different kind of futility case was raised. Imagine that a patient was suffering from pain from disease and asked for pain relief. Imagine further that the physician, without trying anything, refused to give pain medication on the grounds that it was futile, that it would not alleviate the pain which the patient had. I suspect we would all be horrified at such a prospect. If we were asked why, we would respond that one of the purposes of medicine is to relieve pain and suffering and that the physician was wrong not to try to do so. We might understand that the physician could try and fail to accomplish his goals in this case but that not to try was wrong.

I want us to note that we need some conception of a purpose of medicine to explain our outrage at the imaginary physician's behavior. Without that sense of purpose, our response does not make sense. However, with that sense of purpose for the practice of medicine, our response becomes intelligible and the physician's behavior reprehensible. Note that what is not being claimed is that the physician must succeed, only that the point of the enterprise is to attempt to relieve pain and suffering. Note also that I recognize cases at the margin where the patient's pain and suffering has a psychological basis but no physiological one. It also may be that, in the end, nothing which the physician can do will relieve the pain, either for physiological or psychological reasons.

The relevance of this for our discussion should be clear. When the patient demands a treatment which the physician knows will not have the desired physiological effect, the patient either does not understand the limited nature of the purposes of medicine or does not accept the accuracy of the physician's judgment in this particular case. Just as our physician in the previous example did not accept the charge of the practice of medicine, our patient may well not accept the fact that the goals of medicine are limited or may be concerned about the fallible physician.[3] It is not their fallibility which is the issue here but the limited nature of what physicians can do for us.

Another service the religious community might render to the medical community is support for those physicians who argue that the futility debate is an opportunity to open discussions about care for patients who can no longer be cured by medicine [21]. The discussion of caring, long a staple of feminist discourse within health care in general and nursing in particular, ought to receive more support than it does from mainstream theology. It is important to bring to that discussion a richer conception of care than many clinicians hold; some still think of care in terms of pain relief only, instead of a way of accompanying a patient to their death. Here the futility discussion might be one more indication that patients think they have been abandoned by their physicians. If they are going to be abandoned by physicians, some patients will demand that they be killed by those physicians who would abandon them, others will insist that they have a right to refuse what treatments the physicians offer and still others will insist that physicians should do everything for them. If there is one place where theological discussion ought to support voices within secular bioethics, it is on the discussion of the care of patients when cure is no longer possible. That discussion provides an entry point for the other considerations discussed in this essay.

SUMMARY AND CONCLUDING REMARKS

This paper has argued that secular bioethics will not be able to resolve the discussion of futility, given the premises of the debate. Further, secular bioethics, under a guise of neutrality, permits patients to live with illusions about their mortality. Secular bioethics itself is able to sustain only a truncated discussion of the purposes of medicine. Ironically, even the four principles of bioethics are in need of some outside support.

As should be clear by now those secular bioethicists who argue that patients ought to decide whether a treatment is futile are the primary audience for this theological critique. In my view, they permit persons to hold illusionary conceptions of a good life without subjecting those conceptions to critique.

If Christian theologians engage secular bioethicists on the issues I have suggested, there is no certainty that there will be an audience. It is simply too early to predict what will happen.

Religious institutions ought to play a role in these discussions, at least in providing forums for health care personnel to reflect on these issues under other assumptions. In addition, religious communities could begin to display to the larger culture what a medicine with limited purposes might look like. The focus upon cure in modern medicine and the lack of attention to care would be a place to begin that discussion.

One ought to be neither optimistic nor pessimistic, simply hopeful. There are powerful institutional forces which propel medicine to the impasse it finds itself on futility. One can only make suggestions in the hope of finding a receptive ear. If not, it may be that an alternative medicine will have to be sustained and then we will have to turn our backs on technological medicine and the secular bioethics which both shapes and is shaped by it. In my judgment, that time is not yet.

Lafayette College
Easton, PA 18042

NOTES

1. The two cases mentioned earlier hardly constitute a trend in the decisions of courts. There is something, however, worrisome about those decisions. The modern liberal state long has attempted to control all the institutions which might come between the state and the individual. If these initial judicial findings are the beginnings of a trend, it will become harder and harder for physicians to maintain any independence of judgment about what it is that they can and cannot accomplish through medicine. Physicians will fear being overridden by one of the agencies of the modern state. Ironically, as I hope to show below, it will also mean that the liberal state would encourage a context in which we would see ourselves untruthfully.
2. Here is one place where the Christian community might display to the larger world what this might mean. This display may or may not be welcomed; it surely is needed.
3. The difficulties of accepting a fallible medicine could form a whole different essay. Fallible physicians will always be a difficulty for a world in which forgiveness is not practiced. In such a world, physicians are not free to own their errors and patients are not given the opportunity to forgive them.

BIBLIOGRAPHY

1. Beauchamp, T. and Childress, J.: 1989, *Principles of Biomedical Ethics*, 3rd ed., Oxford University Press, New York.
2. Bouma, H. *et al.*: 1989, *Christian Faith and Medical Practice*, Wm. Eerdmans Publishing Co., Grand Rapids, Mich., pp. 57–66.
3. Bresnehan, J.: 1993, 'Medical Futility or the Denial of Death?', *The Cambridge Quarterly of Healthcare Ethics* 2, 213–217.
4. Buehler, D.: 1993, 'Medical Futility', *The Cambridge Quarterly of Healthcare Ethics* 2, 225–227.
5. Cotler, P. and Gregory, D.: 1993, 'Futility: Is Definition the Problem? Part I', *The Cambridge Quarterly of Healthcare Ethics* 2, 219–224.
6. Cranford, R.: 1991, 'Helga Wanglie's Ventilator', *The Hastings Center Report* 21(4), 23–24.

7. Engelhardt, H.T. and Wildes, K.: 1994, 'The Four Principles of Health Care Ethics and Post-Modernity: Why a Libertarian Interpretation is Unavoidable', in R. Gillon (ed.), *Principles of Health Care Ethics*, John Wiley & Sons, New York, pp. 135–148.
8. Fletcher, J.: 1994, 'Baby K, the Supreme Court, and Congress', *Bioethics Matters: A Newsletter for Friends of the Center for Biomedical Ethics* 3(3), 1.
9. Frohock, F.: 1992, *Healing Powers: Alternative Medicine, Spiritual Communities, and the State*, University of Chicago Press, Chicago, Ill.
10. Greenhouse, L.: 1994, 'Court Order to Treat Baby with Partial Brain Prompts Debate on Costs and Ethics', *The New York Times* (February 20, 1994), 20.
11. Hauerwas, S.: 1986, 'Salvation and Health: Why Medicine Needs the Church', in Stanley Hauerwas (ed.), *Suffering Presence*, University of Notre Dame Press, Notre Dame, IN, pp. 63–83.
12. Jecker, N. and Schneiderman, L.: 1993, 'An Ethical Analysis of the Use of "Futility" in the 1992 American Heart Association Guidelines for Cardiopulmonary Resuscitation and Emergency Cardiac Care', *Archives of Internal Medicine* 153, 2195–2199.
13. Johnson, D.: 1993, 'Helga Wanglie Revisited: Medical Futility and the Limits of Autonomy', *The Cambridge Quarterly of Healthcare Ethics* 2, 161–170.
14. Jonsen, A.: 1994, 'Intimations of Futility', *The American Journal of Medicine* 95, 107–109.
15. Kapp, M.: 1994, 'Futile Medical Treatment: A Review of the Ethical Arguments and Legal Holdings', *Journal of General Internal Medicine* 9, 170–177.
16. Lammers, S.: 1993, 'Tragedies and Medical Choices: The Lakeberg Case', *The Christian Century*, 845–846.
17. May, William: 1991, 'Introduction', in *The Patient's Ordeal*, Indiana University Press, Bloomington, IN., pp. 1–14.
18. Murphy, D. and Funucane, T.: 1993, 'New Do-Not-Resuscitate Policies: A First Step in Cost Control', *Archives of Internal Medicine* 153, 1641–1648.
19. Nelson, L. and Nelson, R.: 1992, 'Ethics and the Provision of Futile, Harmful, or Burdensome Treatment to Children', *Critical Care Medicine* 20, 235–239.
20. Rodwin, M.: 1993, *Medicine, Money and Morals: Physicians' Conflicts of Interest*, Oxford University Press, New York.
21. Schneiderman, L. and Faber-Langendoen, K.: 1994, 'Beyond Futility to an Ethic of Care', *The American Journal of Medicine* 96, 110–114.
22. Veatch, R. and Spicer, C.: 1993, 'Futile Care: Physicians Should Not be Allowed to Refuse to Treat', *Health Progress* 74(10), 22–27.
23. Verhey, A.: 1984, 'A Doctor's Oath – and a Christian Swearing It', in D. Smith (ed.), *Respect and Care in Medical Ethics*, University Press of America, Lanham, MD.

MICHAEL M. MENDIOLA

Overworked, but Uncritically Tested:
Human Dignity and the Aid-in-Dying Debate

Throughout the debate and discussion surrounding aid-in-dying whether understood as active euthanasia or assisted suicide,[1] a certain term emerges again and again: "human dignity". Much like the refrain of a well-known tune, we hear repeatedly that we must honor and respect the demands of human dignity in the care of the critically ill and dying, especially those who request the termination of their lives. I have no quarrel with this claim; indeed, I too embrace it wholeheartedly. Yet, as my title is meant to indicate, "human dignity" all too often lies as an under-examined presupposition or background concept in current ethical considerations of aid-in-dying. As one commentator notes, human dignity generally has been "'displayed' and discussed in intellectual abodes, while remaining essentially unexamined and unexplored" ([11], p. 18). In other words, human dignity as a moral concept is most often asserted, as if its meaning and moral weight were self-evident, rather than critically explored; its conceptual substance is assumed, rather than critically examined or justified. It is just such critical, but only preliminary,[2] scrutiny that I undertake in this essay.

I will attend directly to the idea of human dignity, attempting to pinpoint how it is most often understood in secular treatments of aid-in-dying, and proposing an alternative understanding, drawing on Christian ethical resources.[3] Thus, I will attempt to locate how best to understand human dignity and to tease out, if only tentatively, what normative significance it might have to aid-in-dying.

However, it is important to point out a central methodological claim that lies at the heart of this endeavor – it is that one cannot gain an adequate conceptual understanding of human dignity and its moral demands unless one considers the anthropology that lies behind it. By "anthropology", I mean simply our understanding of what it means to be human. In other words, foundational to any proper understanding of the moral demands of human dignity is an understanding of the theological/philosophical anthropology that shapes and informs it. Without such anthropological analysis, "human dignity" remains an empty, highly malleable concept – perhaps of great rhetorical value, but offering little substantive moral guidance. If the concept of human dignity

129

E. E. Shelp (ed.), Secular Bioethics in Theological Perspective, 129–143.
© 1996 Kluwer Academic Publishers. Printed in the Netherlands.

is as central to the issue of aid-in-dying as current usage seems to indicate, then the kind of anthropological analysis that I will undertake in this essay seems requisite.

Before proceeding to the body of the essay, two initial qualifications are in order. First, I will not attempt an ethical analysis or assessment of aid-in-dying directly; my purpose in this essay is not to "take my stand" regarding this issue. My purpose is to deal directly with a background concept that rarely receives direct, explicit consideration. It is my hope that the substance of the essay will have relevance to the issue; however, a head-on ethical assessment of aid-in-dying will not be undertaken in this essay.

Second, as indicated above, I seek to retrieve from Christian sources a foundational anthropology, and allied understanding of human dignity, that is an alternative to that found in most secular bioethical works. However, I cannot claim that all Christian ethicists draw on or utilize that alternative understanding, or are critical of the common understanding found in general, secular analyses of aid-in-dying. Indeed, a comment offered by Leon Kass a number of years ago seems an apt one here. He noted that most religious ethicists in undertaking bioethical analysis today often seem to leave their religious insights at the door, speaking the language of deontology and utilitarianism "just like everybody else". In this sense, they undertake their ethical analyses in a fashion not very different from their philosophical, secular colleagues ([16], pp. 6–7). Hence, my point here is less a descriptive one about current practice among religious ethicists, and more an epistemological one about an alternative, critical vision that can be gained from explicitly religious/ethical resources, even for a purportedly "secular bioethics". Thus, the alternative understanding proposed here is one that I believe to be faithful to Christian insight, and the kind of critical analysis that a Christian should be able to raise (but may not always for various reasons), drawing on explicitly religious perspectives.

I

What understanding of human dignity is generally operative in secular ethical assessments of aid-in-dying? I suggest that a very particular, and faulty construal is evident. At root, much ethical discussion proceeds from an understanding of human dignity as residing in choice and control over the disposition of oneself and one's existential situation. Thus, honoring the demands of human dignity entails allowing people choice and control over whether or not they live or die. A few examples, briefly reviewed, suffice, I believe, to make my point.

In an essay written a number of years ago, Marvin Kohl provides at least the beginnings of an attempt to address the meaning of human dignity directly. In so doing, he evidences the common understanding of human dignity specified above. He proposes that human dignity is best understood in two senses.

Dignity1 connotes the intrinsic worth of human beings – a worth that cannot be lost or diminished, even should one be ill or mentally handicapped. However, sense1 is of little relevance to aid-in-dying, for telling a dying patient that she has dignity is "like consoling the concentration-camp prisoner who is forcefully carried into the 'shower house' by telling him that he is metaphysically free" ([17], p. 133).

In contrast, human dignity2 "connotes having reasonable control over the major and significant aspects of one's life, as well as the ofttimes necessary condition of not being treated disrespectfully". It is this second sense of human dignity that is most crucial to aid-in-dying – for Kohl, the "heart of the matter":

> It has to do with dignity2, the having of which roughly denotes the actual ability of a human being to rationally determine and control his way of life and death and to have this acknowledged and respected by others. This is what is meant when we say that because all human beings have a basic need for dignity they have a corresponding right to be so treated ([17], p. 133).

Here we see dignity expressed clearly and unambiguously as a matter of free choice in the service of self-determining control over one's life and death.

In a more recent work focusing on abortion as well as euthanasia, Ronald Dworkin espouses an understanding of human dignity not dissimilar to that of Kohl (although the fundamental logic of his argument is quite distinct). At root, Dworkin argues that human dignity means respecting the inherent value of our own lives ([8], p. 238). But central to such respect and inherent valuing is self-determining choice and control over the fundamental direction of one's life: "Freedom is the cardinal, absolute requirement of self-respect: no one treats his life as having any intrinsic, objective importance unless he insists on leading that life himself, not being ushered along it by others, no matter how much he loves or respects or fears them" ([8], p. 239). Dworkin offers a strong claim here – one that is open to serious challenge. My aim here, however, is not to engage Dworkin critically regarding this claim; but, again, only to point to the operative notion of human dignity in evidence. As Dworkin states succinctly, "Because we cherish dignity, we insist on freedom . . ." ([8], p. 239).

The final example I take, not from the pages of ethical discourse, but from the world of public policy and statutory proposal: the recent California ballot initiative, *Proposition 161: Physician Assisted Death, Terminal Condition* initiative.[4] Note the following statement from the "Declaration of Purpose" in the initiative:

> Self-determination is the most basic of freedoms. The right to choose to eliminate pain and suffering, and to die with dignity at the time and place of our own choosing when we are terminally ill is an integral part of our right to control our own destinies. . . . The right should include the ability to make a conscious and informed choice to enlist the assistance of the

medical profession in making death as painless, humane, and dignified as possible [22, Sec. 2525.1]. (emphasis added)

Although the ballot initiative may be criticized in any number of its provisions, I cite the initiative in order to highlight the understanding of human dignity that informs it. "To die with dignity" means control over the time and manner of one's death. Human dignity thus entails control over the wide terrain of one's life and destiny. Respecting human dignity demands, not only the negative right of non-interference, but also the positive right of assistance toward self-chosen death by a qualified physician.

These three, all-too-briefly-examined examples lift up what I propose to be a common construal of human dignity in general, secular analyses of aid-in-dying. Other examples are readily available from the secular literature ([2], p. 206), ([23], p. 1381). What all these examples evidence is a common, almost taken-for-granted understanding of human dignity: at the heart of human dignity stands control and choice over one's life, death, destiny. Respect for human dignity entails allowing others such self-determining choices.

II

In the introduction to this essay, I characterized this construal of human dignity as inadequate and faulty. What warrants this assessment; what grounds lead me to it? I will raise two central problems that emerge from the construal of human dignity as choice/control. By far, the most important and weighty is the second; thus it requires close and careful scrutiny.

The first problem is that the understanding of human dignity as choice/control, when applied to death and dying, leads to a good amount of conceptual ambiguity and unclarity relative to who may be said to "die with dignity". If "death and dying with dignity" means a decisive self-control over the possibility, the timing, and the manner of one's death, does that then mean that death and dying without some measure of control or choice is without dignity? This seems, at the very least, a most curious and somewhat problematic implication of the common secular understanding.

For example, my father died a few years ago rather suddenly, after a serious illness from which no one expected him to die. One day he was hospitalized for an acute episode. That evening, an ICU physician informed my sister that my father seemed fine, and stabilized from the earlier crisis that had occasioned the hospitalization. Later, an ICU nurse checked on him; he seemed to be asleep and resting comfortably. When she checked on him a few minutes later, he was dead.

As is clear from this brief narrative, very little choice or control was exercised by my father, or anyone else for that matter, in his dying and death. Yet it seems odd, even a bit presumptuous, to assign his death as lacking in dignity, as he died (to the best of my knowledge) at peace with himself, his loved ones, the world, and God. One must also consider the vast numbers of people who die suddenly, or for whom the issue of choice and control has little applicability or relevance, i.e. "Grandma died peacefully in her sleep".

In light of these terminal situations, the notion of death with dignity as death with control and choice seems to enjoy a very limited range of relevance. Richard Zaner's insightful comments about the experience of renal dialysis seem equally applicable to at least some experiences of critical illness and death:

> The disease, in other words, is intimately experienced as befalling the person, as an unasked-for and unanticipated "happening to me", falling outside of the person's range of possible choices and plans. . . . Thus, the disease, as an affair of embodying experience, is textured by the prominence of accidental circumstance. "It could have been otherwise", is a dimension of the person's experience of his embodiment, but likewise, "I had no choice in the matter", and "why to me (and not someone else)?" ([32], pp. 50–51).

Thus the notion of human dignity as choice/control is very limited and limiting: limited in its range of applicability and limiting in giving normative priority to a select type or manner of dying.

In addition, it seems to grant the blessing of dying with dignity solely to those who choose to exercise control over death and dying, or who have had the opportunity to do so. What of my father, what of countless others – did they die without dignity? Again, it seems highly presumptuous to claim so, without a more full account of who he was as person, his sense of identity and values, the local history of his life, and the social, relational and existential context of his dying and death. Only a moral philosophy which values highly choice and control would construe the nature of human dignity relative to death and dying in such a partial and questionable fashion.

The second problem is related intimately to the first. In many ways, this emphasis on human dignity as choice/control has a deep resemblance to and deep resonance with that moral principle which has been said too often to trump all other concerns in contemporary bioethics: the principle of respect for autonomy ([7], pp. 14–17). Indeed, within the secular treatments of aid-in-dying, "human dignity" seems almost to function as a surrogate term, a code word, for respect for autonomy, in that they are perceived almost identically. For example, in their major and influential work, Beauchamp and Childress argue that what is at stake in respect for autonomy is respect for the self-determining choices of others. Note the very way in which they elaborate the concept of autonomy: it consists essentially in a procedural account, specifying the conditions which allow an autonomous choice ([1], p. 69). Clearly, both my characterization of the usual secular construal of human dignity and the normative specification of autonomy seem to be closely allied here. This is important, for I suggest that in this close alliance, we find an important clue to understanding the construal of human dignity as choice/control.

As I have argued elsewhere, what is conceptually prior in this normative account of autonomy as choice is a particular understanding of the human person. At its core stands the sovereign individual who must be protected from outside, unconsented-to interferences from others [19]. A Hastings Center project on chronic illness articulated this individualistic understanding of the human person in a most helpful manner. In examining the "autonomy paradigm" that has so dominated contemporary bioethics, they propose that a highly individualistic anthropology is operative:

> Thus it is not the notion of autonomy per se that we find inadequate. The problem lies rather in the peculiarly individualistic interpretation often given to the concept of autonomy, in which autonomy means freedom from external limits or constraints. In this perspective, the autonomous self is a disembodied self; it is independent of and prior to its social milieu and its bodily condition ([14], pp. 11–12).

In my judgment, the project's assessment here is keenly accurate, and is also helpful in understanding the emphasis given to a procedural account of autonomy.

Only with a normative presupposition of the sovereignty of the individual does an emphasis on autonomous choice and control over one's life, decisions, and actions make any sense. What is essential in this account is that the individual be free to choose according to her own lights, not the content of what is chosen.[5] Only in this way can the sovereign individual be protected from unwanted interferences by others in controlling the fundamental direction and drift of her life. Thus, the emphasis on autonomous choice which characterizes contemporary bioethics, is actually rooted in a very particular understanding of the human person. A procedural account of autonomy and a highly individualistic anthropology are deeply inter-related and reenforcing.

Through this analysis, an important methodological point becomes clear: underlying this construal of autonomy lies a very particular philsophical anthropology. Childress himself illuminates this point well:

> . . . the principle of respect for autonomy is ambiguous because it focuses on only one aspect of personhood, namely self-determination, and defenders often neglect several other aspects, including our embodiment. A strong case can be made for recognizing a principle of "respect for persons", with respect for their autonomous choices being simply one of its aspects – though perhaps its main aspect. But even then we would have to stress that persons are embodied, social, historical, etc. ([7], p. 13).

In this statement, Childress states clearly that a very selective reading of the human person informs the current account of respect for autonomy. I cite him simply to reemphasize the point that an anthropological principle underlies much contemporary ethical elaboration of autonomy and its moral demands.

I have undertaken this analytic review of the principle of respect for autonomy because I believe it to have direct bearing on the concept of human dignity. In other words, I propose that a similar point is relevant to the understanding of human dignity operative in the aid-in-dying debates. Just as autonomy as choice reflects a certain anthropology, so too does human dignity as choice. Jean-Pierre Wils has raised this methodological point directly regarding the concept human dignity generally. He writes, "Anthropologising duplicates the possible uses of the category of human dignity. From now on its evidence seems bound to the context of whatever anthropology in which it appears" ([31], p. 42). In other words, the concept "human dignity", because it is content-poor, may have multiple meanings, dependent upon and duplicated by various anthropologies. Thus, only by specifying the anthropology which informs it does the concept take on substantive content and moral weight. Hence, as Wils rightly proposes, one cannot deduce moral requirements solely from the concept of human dignity, that is without the "context of case-specific concretions of anthropology" ([31], p. 43).

Thus, a particular philosophical anthropology lies behind the secular understanding of human dignity as residing in choice and control over one's life and death. I propose that it is the very same anthropological perspective that I argued above underlies contemporary bioethical expositions of respect for autonomy: the sovereign individual, surrounded by and protected from others by the primary moral obligation of negative autonomy, non-interference. Flowing from this fundamental vision of the self-determining individual emerges an ethical emphasis on autonomous choice as the "sword and shield" by which one effects control over one's life and self. I take these two symbols directly from the President's Commission's volume *Making Health Care Decisions*. There autonomy is defined in the text and literally illustrated in drawings as a sword and a shield. As a shield, it is meant to protect patients from "arbitrary, albeit often well-meaning, domination by others", assuring the patient's "freedom not to be forced to do something". As a sword, autonomy is an expression of "creative self-agency": "although the reasons for a choice cannot always be defined, decisions are still autonomous if they reflect someone's own purposes rather than external causes unrelated to the person's 'self'" ([21], pp. 45–46).

In sum, as with autonomy, so too with human dignity: both select out a fragment of human life and experience – choice and control – and raise them to an unwarranted ethical prominence and centrality, because of the atomistic, individualistic understanding of the human person which underlies them. Drawing on the symbolization of autonomy noted above, I raise the question: Is human dignity most adequately conceived of as a sword and a shield, for warding off interfering intruders and for cutting a self-determining path of self-agency? I think not, as I hope to make clear below. Perhaps a more adequate anthropology might serve us and the dying better; I turn now to Christian theological sources in the effort to locate just such an alternative.

III

What image or notion of the human person does Christian theological tradition offer to us? It is difficult to locate one such primary conception, since there are many and diverse Christian theologies, not all of which enjoy coherent resolution. Yet, at the risk of over generalizing a rich and diverse theological tradition, I propose two constant themes which define the Christian understanding of the human person, and thus are of central relevance to a Christian understanding of human dignity.

First, Christian theological tradition tends to understand the human person in a highly relational fashion. This relational understanding contains a dual, inter-related thrust. At the heart of this theological tradition stands a central affirmation: the human person stands in a direct, fundamental relationship to God. Helmut Thielicke gives clear articulation to this primary relationship as a central constitutive aspect of being "human". He proposes that the human relationship to the Creator is the "decisive point of humanity":

> We are not just beings that enjoy certain special qualities such as reason, conscience, or an upright stance. Our dignity and inviolability is that we come from the hands of the Creator and that these hands are upon our lives and direct them until they go back to the one from whom they came. The mystery of human life is that the Lord of life summons it to a history with himself. . . . Our worth is based upon this imparted sharing in the divine life. The history with himself to which God has called us constitutes the basis, goal, and meaning of our existence ([27], p. 85).

It is this relationship with the divine which constitutes, for Thielicke, our "alien dignity", the fundamental worth and dignity which human beings enjoy because of this divine invitation to share God's history ([27], p. 85).

Paul Ramsey offers a perspective similar to that of Thielicke in his reflections on a central Christian symbol, the imago dei, i.e. that human beings are in some manner the image of God. In a manner analogous to Thielicke's notion of "alien dignity", Ramsey argues against any notion of the image of God as residing in human capacities, qualities, or faculties. Rather, for Ramsey, the image of God is a dynamic relational attribute; it resides in a "responsive relationship to God". "The image of God is rather to be understood as a relationship within which man sometimes stands, whenever like a mirror he obediently reflects God's will in his life and actions" ([24], p. 255). Again, as in Thielicke, the divine relationship is seen as a fundamental, constitutive dimension of being human.

This theological perspective is also evident in Catholic tradition. Michael J. Schuck, in his analysis of Catholic social teaching, raises a very similar point. He notes that informing this rich and broad body of social ethics is a central theological affirmation: the capacity for an "ongoing, affective relationship with God". It is this relationship which undergirds human dignity – a point stressed again and again by the Second Vatican Council ([26],

p. 211). Thus, only in relationship to God is full humanity and full human dignity manifested ([25], p. 136).

What is crucial to see, in these all too brief theological citations, is that the human person (and thus human dignity) is understood in a profoundly relational way; it is rooted in a relationship with God. This has profound implications for a Christian theological anthropology. What is the essential and decisive qualification is "relationship" – human beings are constituted by relationship, not only to God, but to other human beings as well. Again, Thielicke is helpful at this point. He notes that we can never understand ourselves in isolation, for our identities involve more than isolation. "What we understand by individuality . . . is something that is enmeshed in a system of relations and influenced by this system". But this relational quality even of human individuality is rooted in the relationship to the divine: "I live and move and have my being in a being for others and the world with its infinitely differentiated nexus" ([27], p. 208). Thus, for Thielicke, and I daresay for Christian tradition generally, to be in relationship to God is to be in relationship to the world and others, i.e. "I have my being in a being for others". One cannot in Christian tradition speak of the "vertical relationship" to the divine without at the same time embracing the "horizontal relation" to others and to the world. From a Christian perspective, the vertical is in the horizontal ([18], p. 39). What a different understanding of the human person than the secular understanding explored above!

The second constant theme in a Christian understanding of the human person is that the whole person, in the full dimensions that constitute our humanity, must be considered – as the Second Vatican Council put the matter, "the human person integrally and adequately considered (*personam humanam integre et adequate considerandam.*)"[6] It is the whole person that stands in the relational context with and before God and others in the grace of creation and redemption. Hence, the full dimensions of the human person require explicit consideration.

Richard A. McCormick has undertaken the conceptual task of specifying and giving content to this notion of "the whole person". Drawing on the work of Louis Janssens,[7] he articulates eight dimensions or aspects of the person integrally and adequately considered: human beings are subjects, created for responsible activity in the world; they are bodily subjects, with corporeality as essential to their humanity and dignity; they are part of the material world, invited to agency in the world of nature and culture; human beings are essentially interpersonal, coming to full humanhood in essential interdependence with others; they must live in social groups with structures and institutions appropriate to their humanity; they are called to a transcendental relationship with God; they are historical beings, developing as individuals and in social-cultural groups; human beings, while unique and original, are equal.

I list these various dimensions or aspects, not because they specify the definitive list of the component dimensions of being human, but because they begin to indicate the breadth and richness of perspective that is entailed in considering a human being integrally and adequately. In a manner reminiscent of Eric Cassell,[8] McCormick here raises a direct challenge to, or at least a helpful alternative to, the philosophical anthropology previously discussed. In contrast to a notion of human dignity grounded in a vision of the self-determining individual exercising choice and control, we see here the fundaments of a much broader, and, in my judgment, a much more adequate vision of what it means to be human. Utilizing this alternative anthropology occasions accordingly a shift in the construal of human dignity.

At heart, human dignity cannot reside simply in choosing and controlling the time and manner of our lives and deaths; honoring human dignity must mean more than not interfering in such choices in the name of respecting autonomy. While I do not deny the importance or relevance of self-determining choices in honoring the moral demands of human dignity, such an account of the moral dimensions of human dignity appears thin and flat when contrasted with the theological perspectives analyzed. Human dignity resides in the full dimensions of human being – relational, bodily, transcendental, etc. Honoring the demands of human dignity means a recognition of, and positive efforts to enhance and nurture these multiple, rich dimensions. Surely choice and self-direction are a necessary aspect of any ethical attending to human dignity; but they are never sufficient.

For example, surely I want some freedom and choice about with whom I will relate in work or in leisure; yet, my relationships are not exhausted by the mere possibility or exercise of choice. Friendship clearly is not exhausted existentially or conceptually simply by the fact that I choose my friends. There may also be times in which I am called upon to suffer with a friend, even if it was not exactly "my choice" to do so. We are also born into relational contexts and histories, i.e. families, neighborhoods, and cultural-ethnic groups, which we do not "choose" and over which we have little control. From the alternative anthropology proposed in this essay, giving moral primacy to choice and control presents a meager and distorted view of the human person ([5], p. 175). Thus, it gives a meager and distorted view of human dignity and what is morally entailed in honoring it.

In sum, in light of the alternative anthropology drawn from Christian theological perspectives, I perceive the ethical task to be that of a continual effort to define and refine a conception of human dignity that meets the full requirements of a human being "integrally and adequately considered". I stress the term "continual effort", for clearly I have only indicated very initial, preliminary lines of thought and analysis in this regard.[9] However, having laid this preliminary foundation, in closing I will indicate a few implications of this reconceptualization for the issue of aid-in-dying.

IV

I have argued for a shift, a reconceptualization of the anthropology that under-lies construals of human dignity. I want to suggest briefly by a few examples the importance it may have for the ethics of aid-in-dying.

Like its surrogate, respect for autonomy, human dignity as choice/control gives ethical emphasis to the possibility and exercise of choice – that a person be free to make a free and voluntary choice, and not the moral evaluation of the content of the choice. But as Courtney Campbell argues, ". . . moral attention must necessarily focus on the substance of the request [for aid-in-dying]" and he questions "whether patient self-determination is or should be the exclusive moral consideration in assessment of doctor-assisted suicide or active euthanasia" ([4], p. 130). Thus, he continues, while self-determination [in my terms, human dignity as choice/control] is a necessary consideration, it is never sufficient. Instead, we must distinguish between "having a right" and "right conduct" ([4], p. 131).

A shift toward the alternative anthropology I propose from Christian sources may assist in the move away from the kind of procedural ethics that Campbell criticizes. If we are indeed relational beings, as the Christian anthropology proposes, then we are moral beings in relationship as well. The moral life is not separate from, but is located in the midst of and arises out of that nexus of relationships that constitute the human person. Thus honoring human dignity in this relational sense must entail the risk of interpersonal moral assess-ment, reflection, and discussion. I stress the term "risk", for there is always the danger of heteronomy, of imposing on others moral positions different from their own. Yet, not to engage one another in moral discussion and assess-ment, particularly in such a weighty issue as aid-in-dying, runs an equally great risk of ethical solipsism, a kind of it is-morally-right-because-you-choose-it position.

This effort to engage one another morally as well as pastorally may be of great importance to the critically ill and dying. As Arthur Frank writes,

> After persons receive a diagnosis of serious illness, the support they need varies as widely as humanity itself. . . . It takes time for an ill person to understand her needs. The caregiver cannot simply ask "What do you need?" and expect a coherent reply. A recently diagnosed person's life has already changed in more ways than she can grasp, and changes continue throughout critical illness. Part of what is "critical" is the persistence of change ([10], p. 47).

It is this reality of change and of the difficulty in pinpointing needs that leads Frank to suggest that caregivers be actively involved in helping the critical ill "figure out what he needs" ([10], p. 47). I don't think it too far-fetched to include moral reflection and discussion as something the very sick and dying may need. Again, while the sick and dying must never be forced into such ethical discussion or reflection, equally they should not be denied the opportunity in the name of "autonomous choice".

In this way, true honoring of the dignity of the sick and dying, informed by a relational anthropology, challenges the model of choice and control regarding its sufficiency normatively particularly for the care of the critically ill and dying. However, a renewed notion of the human, drawing on the alternative anthropology, also provides us with a wider lens for seeing the humanity of the sick and the dying. As Eric Cassell has insightfully proposed, human persons can suffer in the various aspects or dimensions of their humanity, i.e. the social, the corporeal, etc. To the degree that medicine fails to take into account the multi-layered dimensions of the human person, to that degree medicine may inflict greater suffering in the course of treatment ([6], p. 639).

Analogously, I argue, a notion of human dignity that fails to incorporate the full dimensions of human personhood is not worthy of the name. In stressing choice and control, it may blind us to the many dimensions of the person that are threatened by serious illness and that require careful fostering and nurturing. In light of the proposed alternative anthropology, the following kinds of factors readily come to mind for the critically ill and dying: the need for adequate pain management ([28], p. 847), the need to address the loneliness and isolation that the dying often experience ([12], p. 168), sensitivity to the institutional context/social setting of dying and its effects on the dying person ([12], [29], pp. 29–64), awareness of gender [20] and racial [15] biases and their impact, and the need for appropriate pastoral care. Again, my effort here is not to imply that self-determining choices are irrelevant or not an important ethical concern. Rather it is to suggest that such an emphasis is simply insufficient, and if given prominent or exclusive attention, may indeed blind us to the full dimensions of human personhood that an adequate concept of human dignity should allow us to see.

In my critique of the "autonomy paradigm" above, I drew upon the work of a Hastings Center project on chronic illness. This project called for a reconceptualization of autonomy. Toward that end the project authors propose a notion of autonomy that is most relevant to the care of the dying and thus to aid-in-dying. They suggest that autonomy be understood not as,

> . . . some a priori property of persons abstractly conceived. It is an achievement of selves who are socially embedded and physically embodied. . . . Autonomy is something that grows out of the physician-patient relationship, not something that presides over it ([14], p. 12).

In like manner, I propose that human dignity is best conceived as socially-embedded and physically-embodied, reflective of the human person integrally and adequately considered. Human dignity is not simply a static quality of human persons, but an achievement that grows out of the effort to recognize, protect, and nurture individuals surely as "choosers", as "decision-makers", but always as something much, much more. I have suggested that an anthropology, drawn from Christian sources might be of assistance in such a reconceptualization. I have attempted to indicate, however preliminarily, the implications of such a reconceptualization for the issue of aid in dying.

Pacific School of Religion
Berkeley, California, U.S.A.

NOTES

1. Although efforts have been made to distinguish these two actions both conceptually and normatively [23], [30], I will treat them as one for purposes of this essay. My central concern in this essay is less with the specific normative assessment of active euthanasia or assisted suicide and more with the exploration and assessment of a background concept which informs discussion of both.
2. A fully adequate exploration of the substantive dimensions of the concept "human dignity" would require a lengthy exposition of its conceptual history, i.e. how the concept has been understood theologically and philosophically across the span of Western ethical reflection. Such a task is clearly beyond the scope of this essay; hence, I refer to my own analysis here as "preliminary". For helpful historical perspectives, see [31].
3. I limit my critical retrieval of human dignity to the Christian tradition. However, my conceptual elaboration of the meaning of human dignity and what normative requirements it entails might have resonance within other religious traditions. However, I have not pursued such issues here.
4. Some may question the validity of including legal language from a ballot initiative as a representative example of contemporary bioethics. After all, the purpose of ballot initiatives is not to provide ethical analysis, nor are such ballot initiatives known for their conceptual clarity or rigor! While acknowledging such cautions, I nonetheless include the ballot language here for two reasons. First, the initiative is important because it could have become law in the State of California, and thus solidified in a legal and very public manner the notion of human dignity utilized in the initiative. Second, as a public document, it expresses a certain understanding of human dignity, and could also form public opinion about the same. In this symbolic function, it could be of greater formative significance than many essays in bioethics.
5. As Daniel Callahan states pointedly, "People in the field have touted and honored the value of autonomy but have been reluctant to talk about how one might appropriately use autonomy. . . . We wish as a society to find consensus, a middle ground, but to find it, if at all possible, with the fewest restrictions on freedom and the least presumption of any deeper theories about the human good, human welfare, and human destiny. We thus often leave to one side what I believe to be the most important moral questions" ([3], p. 57).
6. Cited in ([18], p. 15).
7. See [13].
8. See ([6], pp. 641–643).
9. Underlying my call for examination and analysis of the anthropologies which inform the concept, human dignity, stands a difficult conceptual problem which I have not addressed. Given the historical and social locatedness of all our attempts to specify what it means to be human, how do we judge the relative adequacy of the multitude of possible interpretations? I know of no theoretically "tidy" answer, other than to engage in the effort of drawing out the implications of various interpretations, and testing them in light of what seems to serve human beings well, in various spheres and arenas of human life. This I have attempted to do in this essay, relative to the question of aid-in-dying. I am, however, cognizant of a circularity here. To test various interpretations of what it means to be human in light of

what seems to serve human well-being already seems to presuppose some account of what it means to be human. Otherwise, we would have no criteria by which to judge what "serves human beings well". In other words, any criteria for testing the adequacy of interpretations seems already to presuppose an interpretation, taken as normative.

BIBLIOGRAPHY

1. Beauchamp, T.L. and Childress, J.F.: 1989, *Principles of Biomedical Ethics*, 3rd ed., Oxford University Press, New York.
2. Brock, D.W.: 1993, *Life and Death: Philosophical Essays in Biomedical Ethics*, Cambridge University Press, New York.
3. Callahan, D.: 1988, 'Beyond Individualism: Bioethics and the Common Good', (interview), *Second Opinion* 9, 52–69.
4. Campbell, C.S.: 1992, '"Aid-in-dying" and the Taking of Human Life', *Journal of Medical Ethics* 4, 128–134.
5. Campbell, C.S.: 1992, 'Sovereignty, Stewardship, and the Self: Religious Perspectives on Euthanasia', in R.I. Misbin (ed.), *Euthanasia: The Good of the Patient, The Good of Society*, University Publishing Group, Frederick, Maryland, pp. 165–182.
6. Cassell, E.J.: 1982, 'The Nature of Suffering and the Goals of Medicine', *The New England Journal of Medicine* 306, 639–645.
7. Childress, J.F.: 1990, 'The Place of Autonomy in Bioethics', *Hastings Center Report* 20, 12–17.
8. Dworkin, R.: 1993, *Life's Dominion: An Argument about Abortion, Euthanasia, and Individual Freedom*, Alfred A. Knopf, New York.
9. Dyck, A.: 1975, 'Beneficent Euthanasia and Benemortasia: Alternative Views of Mercy', in M. Kohl (ed.), *Beneficent Euthanasia*, Prometheus Books, Buffalo, pp. 117–129.
10. Frank, A.W.: 1991, *At the Will of the Body*, Houghton Mifflin, Boston.
11. Gaylin, W.: 1984, 'In Defense of the Dignity of Being Human', *Hastings Center Report* 14, 18–22.
12. Glaser, B.G. and Strauss, A.L.: 1968, *Time for Dying*, Aldine Publishing, Chicago.
13. Janssens, L.: 1980, 'Artificial Insemination: Ethical Considerations', *Louvain Studies* 8, 3–29.
14. Jennings, B. *et al.*: 1988, 'Ethical Challenges of Chronic Illness', *Hastings Center Report* 18 (Special Supplement), 1–16.
15. Johnson, P.A. *et al.*: 1993, 'Effect of Race on the Presentation and Management of Patients with Acute Chest Pain', *Annals of Internal Medicine* 18, 593–601.
16. Kass, L.R.: 1990, 'Practicing Ethics: Where's the Action?', *Hastings Center Report* 20, 5–12.
17. Kohl, M.: 1975, 'Voluntary Beneficent Euthanasia', in M. Kohl (ed.), *Beneficent Euthanasia*, Prometheus Books, Buffalo, pp. 130–141.
18. McCormick, R.A.: 1985, *Health and Medicine in the Catholic Tradition*, Crossroad, New York.
19. Mendiola, M.: 1991, *Autonomy, Impartial Rationality, and Public Discourse: A Theological Proposal*, unpublished dissertation, Graduate Theological Union, Berkeley.
20. Miles, S.H. and August, A.: 1990, 'Courts, Gender, and "the Right to Die" ', *Law, Medicine, and Health Care* 18, 85–95.
21. President's Commission for the Study of Ethical Problems in Medicine and Biomedical and Behavioral Research: 1982, *Making Health Care Decisions*, U.S. Government Printing Office, Washington, D.C.
22. 1992, *Proposition 161: Physician Assisted Death, Terminal Condition*, Sacramento, California.
23. Quill, T. *et al.*: 1992, 'Care of the Hopelessly Ill: Proposed Criteria for Physician-Assisted Suicide', *The New England Journal of Medicine* 327, 1380–1384.

24. Ramsey, P.: 1950, *Basic Christian Ethics*, Charles Scribner's Sons, New York.
25. Schuck, M.J.: 1991, *That They Be One: The Social Teaching of the Papal Encyclicals 1740–1989*, Georgetown University Press, Washington, D.C.
26. Second Vatican Council: 1966, 'Pastoral Constitution on the Church in the Modern World (*Gaudium et Spes*)', in W.M. Abbott (ed.), *The Documents of Vatican II*, New Century Publishers, Piscataway, New Jersey.
27. Thielicke, H.: 1984, Bromley, G.W. (trans.), *Being Human . . . Becoming Human*, Doubleday Company, Garden City, New York.
28. Wanzer, S.H. *et al.*: 1989, 'The Physician's Responsibility Toward Hopelessly Ill Patients', *The New England Journal of Medicine* 320, 844–849.
29. Weir, R.F.: 1989, *Abating Treatment with Critically Ill Patients*, Oxford University Press, New York.
30. Weir, R.F.: 1992, 'The Morality of Physician-Assisted Suicide', *Law, Medicine and Health Care* 20, 116–126.
31. Wils, J.P.: 1989, 'The End of "Human Dignity" in Ethics?', in D. Mieth and J. Pohiers (eds.), P. Hillger (trans.), *Ethics in the Natural Sciences*, T. & T. Clark, Edinburgh.
32. Zaner, R.: 1982, 'Chance and Morality: The Dialysis Phenomenon', in V. Kestenbaum (ed.), *The Humanity of the Ill*, University of Tennessee Press, Knoxville, pp. 39–68.

GERALD P. McKENNY

Physician-Assisted Death:
A Pyrrhic Victory for Secular Bioethics

Alasdair MacIntyre begins his famous work *After Virtue* with the claim that the moral discourse of modern western societies consists of mere fragments of once-coherent moralities, and that our moral debates are doomed to insolubility because of the impossibility of attaining coherence in our scattered inheritence [14]. Although MacIntyre may have exaggerated the coherence of our predecessors and failed to appreciate the ability of modern societies to find enough common ground to stumble along, one of his fundamental insights – that many of the traditions we claim have constituted us have in fact become utterly strange and distant – seems basically correct. One need only examine Augustine's treatises on lying, for example, with their thoroughgoing rejection of efforts to balance the duty of truth-telling with other claims such as the avoidance of harm, to judge how alienated contemporary Christians of almost every stripe are from at least one of their moral traditions. The same point can be made about suicide. For if there is one non-controversial point in public discussions of the ethics of suicide in general or physician-assisted death (both suicide and euthanasia) in particular, it is that the traditional arguments against suicide no longer claim a hearing in the public realm.

For those who recognize this situation, there are four plausible strategies. First, one may herald the demise of the traditional arguments as another stage in a continuing narrative of moral progress in which patient self-determination reigns supreme over the kingdom of self-legislating ends. This is increasingly the path taken by secular bioethics and is the target of my criticism of secular bioethics in this essay. Second, one may oppose lifting legal and professional strictures against participation in suicide on the grounds that it would involve negative consequences for third parties (patients would be subtly coerced by burdened families or cost conscious institutions to request assisted suicide), for medicine (assisted suicide would undermine the trust patients place in their doctors), or for society (the value of human life in general would be undermined). One would thus maintain a prohibition against physician-assisted death but with two departures from the traditional prohibition: the prohibition would entail the wrongness not of individual acts but of a policy or practices permitting physician-assisted death, and the prohibition would

145

E. E. Shelp (ed.), *Secular Bioethics in Theological Perspective*, 145–158.
© 1996 *Kluwer Academic Publishers. Printed in the Netherlands.*

be grounded on modern consequentialist considerations rather than on the moral convictions of religious or professional traditions. Such consequential considerations are often invoked in secular bioethics in order to set conditions and safeguards for physician assistance in death rather than to reject the latter altogether. These first two strategies both fall within secular bioethics. The final two come from outside. Hence, third, one may continue to affirm the traditional arguments but, in clear contrast to most of those who formulated them, maintain that their validity is recognizable only within the confines of a community with particular faith commitments or understandings of right and wrong. Assuming there are no additional, public reasons for rejecting physician-assisted death, this kind of claim would be compatible with the acceptance of the latter in law or public policy. Fourth, one may take the demise of the public force of the traditional arguments as an occasion to re-examine ourselves, our communities, and the practice of medicine in a more fundamental way. This strategy would ask why the traditional arguments no longer work for many, what factors in the society and in the practice of medicine have made physician-assisted death the issue it is, and whether secular bioethics unacceptably narrows the range of what is morally significant in physician-assisted death.

I am sympathetic to the third strategy, but in what follows I pursue a limited version of the fourth strategy. Specifically, I will argue that the treatment of physician-assisted death in secular bioethics is the product of a society that lacks both the will and the capacity to examine its responsibilities toward the dying and its practices that have made dying a lonely and meaningless experience. The compassion and sincere moral concern which doubtless motivates secular bioethicists masks an underlying narrowness in analysis and a callousness in actual performance. The narrowness enables secular bioethicists to congratulate themselves for giving suffering patients the choice of an early death without questioning the conditions that lead patients to request such a death or the responsibilities others may have to alleviate those conditions. The callousness enables them to leave all of those conditions in place and thus guarantee that more people will be driven to "choose" such a death. Secular bioethics, in other words, is the perfect fig leaf for a society that lacks the moral courage to question its practices and the moral resources to create alternative practices.

It may surprise the reader to learn that neither I personally nor my argument here rule out physician-assisted death in all cases. In fact, my strategy here will be to forego reliance on traditional and more recent arguments that reject physician-assisted death altogether. Instead I will show how secular bioethics is able to assume the implausibility or at least uncertain character of all such arguments. I will then use the work of two contemporary theologians to show how secular bioethics unacceptably narrows the range of appropriate moral concerns. The expanded range of moral concerns I argue for is different from many of the concerns imbedded in traditional arguments but compatible, I believe, with both traditional and more recent theological positions.

I

I begin, then, by rehearsing some traditional arguments against suicide in order to highlight our distance from our moral traditions. (Since such arguments were given by philosophers such as Plato, Aristotle, Locke, Kant and Sidgwick as well as theologians such as Augustine and Aquinas, the "we" whose traditions they are is quite inclusive, though of course not universal.) My purpose here is to show why a critique of secular bioethics of the kind I am interested in (my fourth type above) will find little help from these traditions. Most of the arguments I discuss are found in Thomas Aquinas ([1], Q. 64 art. 5) but I also draw upon Immanuel Kant and other figures. These arguments are generally relevant to the issue of euthanasia as well. They fall into a history of arguments in the west that hold in common the conviction that suicide is morally prohibited while sometimes (though not in Aquinas's case) recognizing possible exceptions – suicides in order to prevent harm to others or to one's country or to pre-empt an unjust death sentence, for example – in which suicide is permissible. As traditionally construed, the prohibition of suicide includes moral claims referring both to self and to others. I begin with claims regarding the self. Aquinas argues that suicide violates two forms of love to oneself, one grounded in natural law, according to which each being seeks its preservation, the other grounded in charity. Kant also finds in suicide the violation of a duty to self, though he refers not to love of oneself but to one's status as a moral being: an act, he argues, that destroys in oneself the very capacity for moral agency cannot itself be a morally permissible act ([13], pp. 218–220). In many forms of Protestant ethics there is a ground for an additional argument against suicide that leads to the next category. Duties to self in this tradition are often construed as a form of duty to the neighbor: one must care for one's life and well being so as to serve one's neighbor more effectively, or at least to avoid being a burden on others.

I turn now to direct duties to others. Aquinas's second and third arguments involve, respectively, duties of justice to one's community and to God, both of whom have moral claims on the lives of persons. The case of the community involves an important claim that life is naturally ordered such that parts belong to wholes. As such, persons belong to communities as parts belong to the whole; hence suicide injures the whole. With regard to God, suicide for Aquinas violates a biblical divine prerogative to "kill and make alive". Life is God's gift and therefore subject to his power. Taking one's life is analogous to killing a slave and thus sinning against a master or usurping judgment in a matter belonging to another's jurisdiction. The analogy to the master-slave relation was employed by Socrates in Plato's *Phaedo* ([15], 62b–e); it and the more general assumption that God has proprietary rights over human life is a staple of both philosophical and theological arguments.

Before turning to arguments about specific instances of suicide, a final point about the conditions of moral agency is relevant since it too distinguishes traditional from contemporary arguments. For both Aquinas and Kant the prohibition against suicide is supported by natural or moral capacities. According to Aquinas love involves a natural inclination to self-preservation while Kant argues that persons as moral agents can not avoid thinking of themselves as endowed with a freedom that enables them to choose what is morally required.

Aquinas goes on to argue against specific instances of suicide, all three of which are relevant to the issue of physician-assisted suicide. In each case an objection denying the wrongfulness of suicide is shown to rest on an error in moral reasoning or judgement. First is the case of a suicide to escape unhappiness. A common argument on behalf of physician-assisted suicide invokes the extreme physical and emotional discomfort and loss of meaning involved in a debilitating illness. But Aquinas would appear to deny such an appeal. The reasoning is consequential: since death is "the ultimate and most fearsome evil of this life", by taking one's life in order to avoid other evils of life one commits a greater evil in order to avoid a lesser one. Second is suicide to avoid a death that degrades one's nobility. Again, the loss of dignity associated with a protracted dying process is a common argument in favor of physician-assisted suicide. Aquinas addresses the case of Razis in 2 Maccabees 14: 37–46 who, like Saul, committed suicide to avoid an ignoble death at the hands of gentiles. According to Aquinas, Razis misunderstood the virtue of courage. To be unable to bear such evils is to display not courage but weakness of soul. Finally, Aquinas considers suicide for fear of committing a future sin. Some have argued that suicide might be excusable if in the throes of an agonizing death one is tempted to curse God. But Aquinas would appear to reject such an argument since by doing evil in order to avoid an evil (and an uncertain one at that, since it is always possible that God will deliver persons from sins), it violates the fundamental formal principle of morality, which is to do good and avoid evil.

It would be a mistake, of course, to assume that western traditions were unanimous regarding these arguments or even in the general condemnation of suicide. There is good evidence that a relative acceptance of suicide occured not only in Stoicism but in early Judaism and Christianity [8]. But it would be naive to minimize the impact of Augustine's near absolute prohibition of suicide and the force of arguments like those of Aquinas in forming the moral consciences of traditions, including medicine, in the west. Nevertheless, today advocates of the traditional prohibition against suicide are clearly on the defensive. In part this is because fundamental moral and religious beliefs supporting Aquinas's arguments are no longer plausible to many. Among these are theological beliefs. For Aquinas, God's prerogative over life and death is inseparable from his analogy according to which human beings belong to God as slaves belong to their masters. But recent theologians are more likely to conceive of persons as stewards of their lives rather than slaves, and as called to serve God's purposes. This shift is significant for two reasons. First, the reference to divine purposes shifts the emphasis from mere acceptance of God's ownership to determination of the purposes of God, which may leave open the possibility of suicides in accordance with those purposes (e.g. altruistic suicides) and perhaps even when God's purposes for one's life are exhausted. Second, the metaphor of stewardship, unlike that of ownership, implies human consent or participation and therefore calls attention to what factors enable and prevent persons from affirming their stewardship.

The experience of illness and death in a technological society renders other beliefs implausible. The spectacle of a protracted death in an intensive care unit or a long battle with cancer convince many that death is not "the ultimate and most fearsome evil of this life", outweighing all afflictions of body and soul. In this case, the realities of technological medicine may lead one to a different calculation of goods and harms than that assumed by Aquinas. But they may also lead one to different conceptions of virtues. A certain nobility remains in the conviction that courage bears the evil of death at the hands of an enemy. But for many persons no such nobility resides in the conviction that courage must bear the indignities of a debilitating illness which mocks the entire previous life of a person.

Other beliefs which support the traditional prohibition are tied to a natural law framework that describes how God orders human life but that no longer appears to most persons as a deliverance of reason. For Aquinas suicide violates a natural inclination to self-preservation. A certain feature which human nature shares with other living things is regarded as a clue to the divine ordering of life and is thus endowed with moral significance. But others find moral significance in what distinguishes human beings from other creatures. Taking a cue from the Stoics, such persons point to freedom from natural determinants as uniquely human and, as Dietrich Bonhoeffer recognized, find in the capacity to commit suicide an affirmation of what is uniquely human ([4], p. 166). From this perspective, one may draw a conclusion Bonhoeffer rejects and endow this uniquely human capacity with positive moral significance. But even when one does not take this step, the Stoic and Kantian recognition of freedom as a condition of moral action means that suicide can not be prohibited on the grounds that it violates a natural inclination. Hence Kant rules out suicide not because it violates a natural inclination but because it is incompatible with a moral law addressed to human freedom. But the freedom Kant asserted was itself vulnerable to a romanticist rebellion which found the idea of a moral law incompatible with freedom. The result is that in some forms of existential and popular ethics the duty to self is a duty (more aesthetic than moral) to be sincere or authentic in one's moral choices. Suicide can therefore be regarded as morally permissible and even as more desirable than continuing to exist in a radically inauthentic way.

A similar fate befalls the claim that the loss of a part injures the whole community. Even without arguing that the validity of this claim is subject to precise calculations of actual benefits or harms done to communities by the loss of "parts", the claim is counterintuitive in the face of examples of suicides that clearly have prevented injury to the community. One such example is the famous case of Captain Oates who wandered off to a certain death in the cold to prevent the failure of an Antarctic expedition. Other, more prosaic cases of patients who request suicide when their illness has driven loved ones far beyond their capacity to cope with the stress of caregiving or when the cost of the illness imposes severe economic burdens on other family members, depriving children, for example, of a college education, may also be examples.

Finally, and perhaps most significantly, traditional views of the moral accountability of those who consider or commit suicide have become implausible even to many of those who continue to uphold the prohibition on traditional grounds. Among many, this raises questions about the appropriateness of ascribing blame to those who commit suicide.

The questions I have raised regarding the traditional arguments against suicide indicate the difficulties any effort to gain public agreement on a prohibition of physician-assisted death is likely to face. This does not, of course, constitute a refutation of these arguments. But it does indicate that such arguments will be persuasive only to those with roughly the same religious beliefs, moral dispositions, and orderings of goods that, say, Thomas Aquinas adhered to. For my purposes, this means that the critique of secular bioethics I wish to carry out will not be able to rely upon these beliefs, dispositions and orderings.

II

Secular bioethicists appear to be slowly building a consensus according to which the primary morally significant characteristic of physician-assisted suicide and euthanasia is the presence or absence of self-determination on the part of the patient. In other words, if the conditions for self-determination are met (decision-making capacity, absence of depression or temporary rage or frustration, etc.) and the physician consents, then an act of physician-assisted death is presumably justifiable. Most secular arguments of this sort also make room for procedural safeguards and other restrictions. But the justificatory work done by the value of self-determination is central if not exclusive.

I have referred to a consensus, but of course there are some who interpret the progression from the right to reject useless and unwanted medical treatment to the moral permissibility of acts of physician-assisted killing – a progression whose curve very nearly defines the past twenty years of bioethics – as a slide down a slippery slope to the abandonment of the commitment to human life that medicine and indeed western society allegedly stand for. I suspect, however, that most secular bioethicists endorse K. Danner Clouser's more optimistic narrative. Clouser views this progression "as a gradual recognition, the gradual discovery, of one's right of self-determination over against tradition, law, and technology. At each stage, the traditional emotions attaching to that action had to be overcome in the course of reasserting self-determination, and so on to the next stage" ([7], p. 307). In any case, with self-determination generally enshrined as the sole or primary morally relevant feature of individual acts of physician-assisted death, the question is now shifting to whether consequences of its practice should prevent physician-assisted death, justifiable though particular cases of it may be, from being morally acceptable as a practice or a policy.

But before I discuss whether consequences can override the priority of self-determination, we must recall that the value of self-determination has had to overcome resistance to physician-assisted death on the grounds that the latter, in distinction from withholding or withdrawing treatment, constitutes killing the patient rather than letting her die or letting the disease kill him. This obstacle – the prohibition of intentional killing – is the one obstacle that remains after the traditional arguments discussed above have lost their force. Can a significant critique of secular bioethics be launched from this site? Among secular bioethicists, Daniel Callahan [6] and Susan Wolf [16] have attempted to oppose one or another form of physician-assisted death by distinguishing killing from letting die. I am not competent to judge Susan Wolf's claim that the legal permissibility of refusal and abatement of treatment has depended upon the continuing legal prohibition of euthanasia so that acceptance of the latter would jeopardize the legal status of the former. Nevertheless it seems clear that the view that stopping life support could be distinguished from killing in the eyes of many persons was a major factor in the acceptance of the moral permissibility of the former. By the same token, if the distinction between killing and allowing to die were found to be irrelevant in at least some acts of stopping life support, then their permissibility might be able to be extended to acts of physician-assisted death.

There is no need to repeat here the well-known arguments against certain formulations of the distinctions between withholding and withdrawing life-sustaining treatment or between killing and letting die. I will only repeat those of their conclusions that have gained wide acceptance and that relate to the issue I am considering. First, the distinction between withholding and withdrawing treatment is conceptually unclear in some cases and the distinction between act and omission that underlies is unclear in such cases and morally irrelevant in other cases ([3], pp. 196–200). Second, the distinction between killing and letting die rests in many cases on a mistaken view of causality. In Dan Brock's example, the morally relevant distinction between the act of a physician who removes a respirator from a patient at her request and the act of her greedy son who removes it in order to claim his inheritance is not whether the disease killed the person (in both cases the act of removing the respirator killed the person) or whether death is intended (in both cases it is), but in other features of the two acts: the motives of each agent, the consent of the patient, the social sanction for the act of the physician ([5], p. 13; cf. [3], pp. 219–225). One who accepts this line of argument need not believe that these distinctions are always irrelevant and may continue to affirm that in some cases death is not intended simply by removing a respirator from a patient – cases, for example, when the patient would die in roughly the same amount of time whether artificially ventilated or not. But in many and perhaps most cases, the morally relevant issue would be whether or not the killing is justified, not whether the act is an act of killing or letting die. And if one believes that removal of a respirator by a physician from a consenting patient whose death would otherwise have been delayed can be justified, then there seems to be no reason why other forms of intentional killing under similar conditions and in similar circumstances can not be justified.

For purposes of this essay I will not evaluate this line of argument or ponder whether there are other morally relevant features to distinguish some acts of abatement of treatment from suicide and euthanasia, which I believe there are. My point is only that with the general recognition that stopping life support constitutes intentional killing, the last remaining obstacle to the permissibility of individual acts of physician-assisted death has been removed. This resolves the strictly moral issue in favor of permitting physician-assisted death and leaves two nonmoral issues, one prudential and one consequential, unresolved. The prudential issue concerns questions about whether and when patients are making a self-determining choice and who is capable of judging it. Should those who make such requests be presumed to be clinically depressed? Should psychiatric evaluations be conducted before the physician assists in a patient's death? Should the patient be required to repeat a request for assistance in dying on multiple occasions over a certain period of time? These questions indicate that procedural safeguards may be needed to guarantee to the extent possible that requests are genuinely self-determining.

These prudential questions already merge into the issue of consequences. Even if individual acts of physician-assisted suicide are justifiable, would physician-assisted death as an accepted practice or policy lead to ill consequences that could override its justifiability? Three sets of questions are especially relevant. The first, which relates to the prudential issues surrounding self-determination, is whether the general acceptability of physician-assisted death would encourage its extension to those who do not want or request it. Will pressure from burdened families or institutions coerce some to request it even though they do not desire it? Will societal protection of those whose quality of life is judged too low or whose cost to society is judged too high erode if physician-assisted death is a live option for them? The second set of questions asks whether the societal commitment to providing optimal care for the dying or devoting resources to the development of new forms of pain relief will erode if dying patients can simply opt for suicide or euthanasia. The third set of questions asks whether the trust of the public in the medical profession or the trust of individual patients in their physicians will erode if doctors are perceived as those who assist in death.

These consequential considerations deserve a lengthy discussion, but for my purposes it is enough to note two points. First, as with all consequentialist arguments the immediate and certain consequences take priority over the remote and uncertain consequences. Second, self-determination and relief of suffering rank high in the value system of secular bioethics as well as being both immediate and certain, while the consequences mentioned above are all uncertain and somewhat remote. Moreover, as Brock argues, the very grounds on which physician-assisted death is permitted – i.e. self-determination – should itself prevent at least the first set of ill consequences (extension to those who do not want or request it) from occuring given appropriate procedural safeguards ([5], pp. 20 ff.). It seems quite clear, then, that for most secular bioethicists there are no serious moral hindrances to the permissibility of physician-assisted death. Recognition of self-determination combined with procedural safeguards (such as full information, repeated requests, and consideration of all alternatives) and other restrictions (such as requiring that the patient be in a terminal condition) to reduce the likelihood of the most serious consequences seems to cover what is morally at stake in the issue of physician-assisted death. On what grounds would one mount a challenge to this growing consensus?

III

Secular bioethics, like secular moral theories more generally, is narrowly focused on acts and consequences, a focus that, as the previous section indicates, holds for the treatment of physician-assisted death. Criticism, therefore, could focus on the lack of any discussion in secular bioethics of what makes some self-determining choices better or worse than others; in other words, what ideals of a worthy life and good death should determine decisions about the end of life.

I am convinced that such an argument needs to be made, but here I will arrive at its conclusion indirectly through a different criticism. This latter criticism concerns the moral responsibilities caregivers and others have toward those who request assistance in dying. It is clear that in secular bioethics these responsibilities are restricted to the procedural safeguards or conditions discussed above which insure the genuineness of self-determination. But the question secular bioethicists avoid is whether, in addition to this duty regarding self-determination, professional caregivers, families and others (including even the patient himself or herself) have any moral responsibility for the conditions which drive persons to consider suicide and eventually to request assistance in dying. Are there certain conditions under which persons at the end of life lose the capacity to experience their lives as worthwhile and meaningful? Do they and others have responsibilities to avoid those conditions and to foster the opposite conditions that enable them to affirm their lives? More pointedly, are the medical and societal conditions under which persons face death today especially conducive to the loss of senses of meaning and worth? Are there responsibilities regarding these conditions?

An entire bioethical agenda regarding physician-assisted death is implied in these questions: If there are such responsibilities, what are they and how should we regard them? Are they obligations binding upon all and subject only to occasions in which to exercise them, or are they duties of station? How would recognition of these responsibilities change the methods and emphases in our care for dying patients? How do our institutions and practices facilitate or limit fulfilment of these responsibilities? How should we respond to the inevitable failures and limitations in carrying out these responsibilities? If these questions are appropriate, then the silence of secular bioethics regarding them is morally suspect since it fails to raise legitimate questions of moral accountability.

Of course, secular bioethicists do not ignore entirely the question of the conditions that prompt patients to request assistance in dying. Most of them agree that physician-assisted death should be a last resort, to be carried out, if at all, only after efforts to improve the quality of life of the dying patient have failed. The problem is that in the account secular bioethicists usually give, such efforts have the same status as the requirements to insure that patients are fully informed and not depressed, and that the request is an enduring one ([5], p. 20). In other words, all of these efforts facilitate self-determination, preservation of which is the sole moral responsibility. This means that the consideration given to the questions I raised in the previous paragraph is extremely limited and the range of accountability for such factors is extremely narrow.

Interestingly, questions regarding the conditions under which persons are driven to "choose" suicide are the primary preoccupation of recent Protestant thinkers as diverse as Karl Barth, James Gustafson, and Stanley Hauerwas (at least in his early work, from which I draw here). Their accounts of what drives people to consider suicide differ. For Barth it is a deep sense of abandonment accompanied by a desperate grasp at control that Barth interprets in religious terms ([2], pp. 403–407). For Gustafson it is a disorder in the complex relations of interdependence that sustain human life or in psychological and spiritual conditions of life ([10], pp. 198–207). For Hauerwas it is a breakdown in the bonds of trust and care that support human life ([11], pp. 112–114). Since Barth interprets not only the conditions that drive persons to commit suicide but also the overcoming of those conditions in explicitly religious terms, as occuring exclusively between the person and God, rather than in moral terms, his account will not be significant for what follows. But for Gustafson explicitly and for Hauerwas implicitly, the account of the conditions that drive persons to commit suicide entail a reordering of moral priorities. The primary moral consideration is no longer the presence or absence of genuine self-determination but rather the responsibility of persons linked to each other in various ways to maintain the conditions that enable themselves and other persons to affirm their lives as worthy and meaningful. What is significant here is the deferral of the question of the justifiability of suicide as an act – a deferral that enables Gustafson and Hauerwas to explore a moral terrain which standard bioethics overlooks.

The difference I am sketching between secular bioethics on the one hand and theological perspectives such as those of Gustafson and Hauerwas on the other hand is rooted in part in different descriptions of human nature and different conceptions of the good that are connected with these descriptions. I mentioned above that I would raise the question of the good as a criterion for choice indirectly through the question of duties. Now it is clear that both are part of a broader conception of human life as interdependent with others in various relationships or in a community. For Gustafson and Hauerwas, persons are by nature such that their survival and well being depend, among other things, upon such relations of interdependence. For Gustafson, these relations are biological, social and familial; persons exist in complex relations to various wholes. For Hauerwas, they occur in particular communities. Hence for Gustafson, persons have moral obligations or duties to sustain these relations of interdependence in accordance with the possibilities and limitations inherent in the particular kind of relation involved, the occasions for exercising such obligations and duties, and the abilities and limitations of the person who exercises them ([10], pp. 207–213). Human beings must consent to suicides which occur when the conditions for life supported by these relations are tragically lacking ([10], p. 209). Connected with the description of human nature as interdependently existing in part-whole relations is a notion of the good that enables Gustafson to determine when suicide may be morally justified, namely, when all possibilities for contributing to the good of oneself or of the larger wholes have been exhausted ([10], pp. 213–215). Hence a certain description of human nature is linked to a broad understanding of moral duties or obligations which is lacking in secular bioethics, as well as to an understanding of the individual and common good – also lacking in secular bioethics – that may justify suicide on some occasions.

According to Hauerwas, communities are sustained by relations of trust and care that constitute the conditions for the lives of individual persons as well ([11], pp. 65–67, 109–112). However, Hauerwas's emphasis is different from that of Gustafson. The duties or obligations to uphold these relations are essential to his analysis but are not elaborated. Instead, Hauerwas focuses on the obligation of persons considering suicide to keep faith with those relations of trust and care. But he emphasizes that suicide is usually and primarily the failure of the community to supply the conditions of trust and care, not the failure of the individual to keep faith with them ([11], pp. 112–114), ([12], pp. 105–107). Communities, however, not only foster the trust and care that sustain life, they also supply the notions of the good that determine what is to count as a suicide and when it is no longer appropriate to deter death ([11], pp. 114–115), ([12], p. 106). Once again, therefore, a description of human nature is linked to a set of duties or obligations and to a view of the good that determines how life is to be lived to its end, both of which are lacking in secular bioethics.

At this point secular bioethicists may object. They may point out that there are as many versions of these duties and of these goods as there are theologians and philosophers. And here they would be right. They may then go on to argue that they are simply trying to devise a limited moral framework that avoids all of these disagreements. But here they would most likely be wrong.

For the fact is that most secular bioethicists assume their own implicit description of human nature according to which persons are rational individuals in essence, for whom particular views of the good and relations to others are accidents. Only under such an assumption can they assume that self-determination is a core moral value that can be abstracted from, and is compatible with, any view of the good or any relations to others one might or might not have. Since the latter are therefore incidental to a moral evaluation, secular bioethicists can rest content convinced that their morality, while minimal, nevertheless exhausts what a patient who considers suicide can morally demand of others. If one probes further, one will note the way in which the description of the person implied by secular bioethics perfectly mirrors the practices that have made dying such a lonely and meaningless experience for so many persons. For the secular bioethicist assumes a person without family and other relationships, abandoned into the hands of the physician who, as a stranger, can and ought only to insure the patient's self-determination. By assuming rather than criticizing the abandonment of the ill and the absence of anything other than a private and unsharable moral framework within which to understand dying, secular bioethics mirrors the contemporary context of dying; by leading its audience to believe that this context can be made morally healthy by physician assistance in dying, secular bioethics legitimizes it. That secular bioethics should mirror and legitimize contemporary practices of dying and caring for the dying is not surprising. Secular bioethics, after all, owes its authority and its very existence to the moral world created by modern medicine.

The only kind of secular bioethics that avoids these problems is one which recognizes the lack of agreement on any description of human nature, including that of secular bioethics, and thus argues for self-determination not as a moral value but as a side constraint [9]. This form of secular bioethics, as proposed by H. Tristram Engelhardt, would permit physician-assisted death when it is conducted with the permission of both parties, assuming certain conditions and safeguards to insure that it occurs only when such permission is given. Since I believe that, once secular bioethics is found to be one more purveyor of controversial beliefs and convictions, conversations among adherents of various views of the good can be fruitful in bringing to the forefront of moral debate the broader range of questions I have tried to identify here, I do not share Engelhardt's position entirely. Nevertheless, the strengths of his position over other forms of secular bioethics are clear. A secular bioethics of his sort carries with it an appropriate modesty insofar as it recognizes that its claims are incapable of addressing most of what is morally at stake in physician-assisted death. It neither mirrors nor legitimizes contemporary practices but understands the task of creating any such practices – those of secular cosmopolitan moderns that are enshrined in so much of contemporary medicine and those of particular communities – as the task of those who adhere to the beliefs about human nature, the moral convictions, and the institutional arrangements that underlie that practice.

It is only a matter of time before secular bioethics cuts through the last points of resistance and its moral vision is enshrined in public policy and medical practice. But for many who struggle through the agony of the brutal endgame that the process of dying has become for many persons, the victory of their self-determination over the forces of tradition and law heralded by Clouser will be a Pyrrhic victory. For it will not have raised one word in criticism of the ways in which we as individuals and as a society force upon the dying a lonely and meaningless death.

By contrast, theological positions which address the kinds of concerns identified by Gustafson and Hauerwas, if fully developed, would begin not with self-determination but with descriptions of the duties various persons have to care for themselves and others and the institutions and practices related to dying that best facilitate fulfilment of these duties. In the context of this kind of care, such positions would also describe how persons are formed in the virtues needed to make the best choices at the end of life. Moreover, they would also recognize the limitations and failures of care and would indicate what responses, perhaps including physician-assisted death and perhaps not, are appropriate in such circumstances. Finally, these positions would be developed in the ongoing traditions of particular communities and would enliven bioethics with vigorous debates and conversations among the adherents of these various traditions.

Rice University
Houston, Texas, U.S.A.

BIBLIOGRAPHY

1. Aquinas, T.: *Summa Theologiae*.
2. Barth, K.: 1961, *Church Dogmatics* Volume III, Part 4, T & T Clark, Edinburgh.
3. Beauchamp, T. and Childress, J.: 1994, *Principles of Biomedical Ethics*, 4th edition, Oxford University Press, New York.
4. Bonhoeffer, D.: 1955, *Ethics*, Macmillan, New York.
5. Brock, D.: 1992, 'Voluntary Active Euthanasia', *Hastings Center Report,* March-April, 10–22.
6. Callahan, D.: 1989, 'Can We Return Death to Disease?', *Hastings Center Report*, January-February, Special Supplement, 4–6.
7. Clouser, K.D.: 1991,'The Challenge for Future Debate on Euthanasia', *Journal of Pain and Symptom Management* 6, 306–311.
8. Droge, A.J. and Tabor, J.D.: 1992, *A Noble Death*, HarperCollins, San Francisco.
9. Engelhardt, H.T.: 1995, *Foundations of Bioethics*, 2nd edition, Oxford University Press, New York.
10. Gustafson, J.M.: 1984, *Ethics from a Theocentric Perspective*, vol. 2: *Ethics and Theology*, University of Chicago Press, Chicago.
11. Hauerwas, S.: 1977, *Truthfulness and Tragedy*, University of Notre Dame Press, Notre Dame, Indiana.
12. Hauerwas, S.: 1986, *Suffering Presence*, University of Notre Dame Press, Notre Dame, Indiana.

13. Kant, I.: 1991, *The Metaphysics of Morals*, Cambridge University Press, New York.
14. MacIntyre, A.: 1984, *After Virtue*, University of Notre Dame Press, Notre Dame, Indiana.
15. Plato: *Phaedo*.
16. Wolf, S.M.: 1989, 'Holding the Line on Euthanasia', *Hastings Center Report*, January-February, Special Supplement, pp. 13–15.

PAUL D. SIMMONS

The Narrative Ethics of Stanley Hauerwas:
A Question of Method

The emergence of "narrative ethics" has generated considerable excitement and interest. The emphasis on story has instant appeal to an image-oriented generation. Narrative invokes and involves imagination and fantasy as modes of moral understanding. It builds on the insight that the mind is a picture gallery, not a debating hall, as MacNeille Dixon reminded us. A narrative approach seems also to have promise for getting beyond the impasse created by what has been called the "context vs. principle debate" ([1], ch. 3.), or the endless dispute over the status and justification of moral rules.

Narrative ethics involves a gestalt switch or a paradigm shift in the conceptual framework for Christian ethical thought. It is interested in the phenomenology of morality as related to the story or context of which it is a part. Story is basic, theories and processes are derivative. The formation of character as well as the shape of moral discourse, central paradigms, images and metaphors depend upon the narrative to which they belong.

In spite of its excitement and promise, narrative ethics has generated considerable opposition and criticism. A major part of the problem is the tremendous variety among exponents, both as to the meaning of "story" or narrative, how story shapes morality, and what moral commitments become normative. James McClendon [14] and Stanley Hauerwas mean very different things and reveal quite different attitudes at significant points. Even so, they quote one another, express appreciation for one another, and share certain commitments and moral perspectives on topics such as non-violence.

THE PROBLEM OF METHOD

I propose to deal with the question of method in the thought of Stanley Hauerwas who is, if not the most prominent, certainly the most provocative of the Christian narrative ethicists. Few narrativists have become so extensively engaged in the public policy debates concerning medical ethics. Hauerwas has done so with such profound convictions and emphatic pronouncements concerning "the Christian" perspective, that dialogue with secular

159

E. E. Shelp (ed.), Secular Bioethics in Theological Perspective, 159–177.
© 1996 *Kluwer Academic Publishers. Printed in the Netherlands.*

bioethicists or even other Christian ethicists has been difficult if not impossible. Some light might be shed on his approach that will contribute to understanding his methodological assumptions and thus open avenues for discussion and possible dialogue.

My analysis/response starts with an issue in biomedical ethics and examines the way in which Hauerwas deals with the subject. I am aware that this reverses the flow of thought he employs in doing narrative ethics. To start by stating a problem he regards as the approach of "quandary ethics"[1] ([4], pp. 16 ff.), ([12], pp. 56 ff.). Even so, it will have the advantage of concentrating the mind in such a way as to make possible an examination of the assumptions that determine Hauerwas' posture on an issue which is of enormous importance. The issue is abortion.

What is of interest is not so much the fact that Hauerwas strongly opposes the practice[2] [6], but how he approaches it. My aim is not to argue pro or con about abortion, but to isolate variables which are crucial to Hauerwas' narrative method. My thesis is that his attitude toward abortion is shaped by certain assumptions which are highly problematic, but that are basic to his construal of "the Christian story". The narrative he would have shape the Christian perspective toward abortion is not so much drawn from Scripture as it is from rather eccentric practices and beliefs discernible in various Christian traditions. The problem is therefore not with narrative as a method of doing ethics[3] but with Hauerwas' personal construal of "the Christian story". Narrative ethics need not result in the harsh rejection of abortion that Hauerwas advocates. Examining just why he argues as he does might therefore serve to clarify underlying issues that explain his strong feelings on this topic.

A study of method in ethics involves an examination of the pivotal presuppositions or variables that determine the outcome of moral argument ([15], p. 57). Hauerwas' method, for instance, involves five assumptions of critical importance for his approach to abortion. These are: (1) the narrative structure of Christian existence in the world; (2) the church as a community of character; (3) attitudes toward sex and childbearing; (4) sanctification in the Christian life; and (5) tragedy and suffering as signs of God's kingdom in the world. These are interrelated and interdependent themes in his writings, of course. Understanding how he construes each of these as essential to "the Christian story" is to understand and explain his posture on abortion.

NARRATIVE AS CLUE TO CHARACTER

The first clue to his thought is his particular interpretation of narrative as the clue to Christian existence. Character is the central concern of morality since it is the name we give to the cumulative source of human actions ([4], p. 29) and is inseparable from the narratives which develop it. Hauerwas does not so much tell stories, as appeal to what he calls *the Christian story.*

He seems not so much interested in the ways that various "stories" (whether from Scripture or history) might function to influence moral acts. He says that his emphasis on narrative is not the central focus of his position, since it "is but a concept that helps clarify the interrelation between the various themes I have sought to develop in the attempt to give a constructive account of the Christian moral life" ([7], p. xxv). He holds a conceptual image or imaginative construct of the story of Jesus that provides the framework for dealing with character, vision and virtue in the life of the believer. How these are related and the effect they have on the moral life seems to be what he means by "narrative structure".

The shift from "stories" to "story" enables Hauerwas to make two moves that are critical for his moral thought. First, it makes possible a focus on "character" instead of "decisions" based on "facts" or data more or less rationally and/or scientifically gathered. "It is character", he says, "inasmuch as it is displayed by a narrative, that provides the context necessary to pose the terms of a decision, or *to determine whether a decision should be made at all*" ([4], p. 20). The use of violence, for instance, is not a matter of a decision for those who are non-violent, any more "than the courageous can decide . . . to be cowardly" ([7], p. 123). The facts are not as crucial as "are the stories that provide us with the skills to use certain moral notions" since our moral perceptions are narrative dependent ([4], p. 22).

The second value of this use of "narrative" is that it enables Hauerwas to account for subjectivity and commitments in moral belief and behavior. He openly embraces the bias that belongs to "subjective beliefs, wants and stories of the (moral) agent" ([4], p. 16), and thus establishes a frame of reference for dealing with truth that is highly particularistic. The assent of faith is prior to and far more important than the structure of reason or pattern of rational analysis that might be involved. Story is prior to explanation as a vehicle for understanding and action.

The shift of focus from "issue" to character or virtue is a critical move since it insulates moral perceptions and commitments from rational analyses and criticism by other points of view. Thus the abortion issue is not a matter of reasoning from a notion of "the sanctity of life", or a principle against willful and deliberate killing, any more than objections to slavery are a matter of reasoning on the basis of personhood. Both are regarded as contradictory to the narrative structure of existence to which Christians belong ([7], p. 118). Thus, abortion is not a question of whether the fetus is a person, but a question about the kind of person *we* are. Christians are a people who do not engage in violence or coercion since we are fashioned by a story that releases us from such destructive alternatives ([4], p. 35).

The problem of method seems immediately apparent since what Hauerwas calls character sounds very much like what others would define as a principlist or even a rules-oriented approach. He states it strongly as being an issue of the "connection between the stories that form our lives and the prohibitions and positive commitments correlative to those stories" ([7], p. 118).

For him, there are basic moral postures that belong to the narrative structure of Christian being in the world. Being part of the story means that we "in fact [are] assuming a set of practices which will shape the ways we relate to our world and destiny" ([4], p. 36).

Hauerwas is openly confessional as a Christian theologian. He insists on the particularity of the story "adopted" or the commitments of faith. Like James McClendon, he identifies with certain commitments within the Christian story. McClendon is clearer about the distinctive stories he accepts, however, openly identifying with "the baptist story" of the Christian life, which is rooted in the Radical Reformation and its commitment to pacifism or peacemaking ([12], p. 19). Hauerwas is less specific about such details but quite emphatic in insisting that his is the biblical story of Jesus which enables us to speak of God and our relation to him ([4], p. 79). The difference in the two might be stated thus: McClendon is a Christian who is committed to the baptist understanding of Scripture and the nature of faith. He is thus open to accepting the fact that other Christians will accept different stories and interpret Scripture from different points of view. There is a pluralism within Christianity that makes it necessary both to declare and to examine the stories that shape our faith.

Hauerwas admits pluralism but questions its validity. His construal of the Christian story leaves little room for disagreement or variety in the witness of faith. But the assumptions or images that constitute his construct of that story are not clearly stated. He is emphatic in saying that there are require-ments and prohibitions that are basic to being Christian, that is, to having one's character shaped by the narrative of Jesus, or having one's being in the world ([4], p. 73). He is not clear, however, about how such norms are constitutive of the story of Jesus and just how these define Christian character.

Hauerwas assumes that there is *one* reading of the biblical story that yields norms ("requirements and prohibitions") which are incumbent upon all Christians. He thus writes with a fervor on certain topics which seems to betray some of the basic assumptions of the method to which he is committed, that is, what he means by the narrative structure of ethics. What might be construed as deontological rules and principles are given meaning and grounding in a particular context, namely, the community which fashions character.

THE CHURCH AS MORAL COMMUNITY

For instance, the story of Jesus is also the story of a distinctive community, the church. Ecclesiology is crucial to Hauerwas' method. He emphasizes the nature of the church as a moral community that forms faithful persons. Behind and basic to all that Hauerwas says about character, morality and narrative structure is a certain understanding of the church and its role in the forma-tion of character.

The community is prior to the individual as the locus of God's transfor-mative activity, he says ([9], p. 11). Questions about the truth of Christian convictions cannot be separated from their ecclesial context. As he says, "the form and substance of a community is narrative dependent and therefore what counts as 'social ethics' is a correlative of the content of that narrative" ([6], p. 20).

Hauerwas is rightly concerned about the church's identity in the world which is tied to his understanding of the story of Christ. Being distinctive means both embodying the identifying characteristics of the foundational story and being distinctive from the world.

The church's task requires it to stand in sharp contrast with the world. It is to provide unequivocal moral guidance by living faithfully to the story by which it understands itself and the world. Its first moral task is to initiate people into the story in a manner that forms and shapes their lives in a decisive and distinctive way ([9], p. 20). Community norms are to be taught and inculcated as the nature and content of the virtues which define Christian living.

Further, the church has various strategies for enforcing conformity or obedience to its moral norms or behavioral expectations. The task is not so much to convince or persuade the world or those who disagree with its teachings or preachments, ". . . but rather to witness to a God that requires confrontation" ([5], p. 73). Images of discipline – perhaps even of being banned from the Lord's table or from the fellowship of the faithful – pervade Hauerwas' writings. Those who do not live the story (as he defines it) show that they are "outside" the church. As he says, "Christians can be found on nearly every side of any issue, but they may not be found there as Christians" ([5], p. 74).

Such strong language underscores the nature of the church as a moral community and establishes its educative role as that of fashioning a "faithful" people. Rigidity is combined with generosity, however, in the second moral role of the church, which is to provide support and sustenance for those facing tragedy. Whether the member is facing a problem pregnancy, the birth of a handicapped child, or economic distress, the church should be a community of shared burdens: "The church, in its finest expression, is the gathering of a people who are able to sustain one another through the inevitable tragedies of our lives" ([5], p. 73).

The image of the church that emerges from such statements reflects something like the communitarian ([11], p. 9) motif found in such traditions as Mennonism, Roman Catholicism, and Methodism. Each of these have influenced his thought and are discernible in his development as a Christian theologian. A serious – even stern and harsh – approach to certain moral practices characterizes each tradition, or at least a particular strand within these traditions. He goes so far as to argue that being part of the Christian community means accepting certain prohibitions, which are "the markers for the outer limits of the communal self-understandings". Christians are part of a community that practices virtues, he says ([7], p. 76).

Failing to embody these virtues, i.e. violating these practices, "means that one is no longer leading one's life in terms of the narrative that forms the community's understanding of its basic purpose" ([7], p. 119). In other words, whether one belongs to the community of faith or not is determined by adherence to norms which Hauerwas regards as self-defining and non-negotiable. The church is constantly to test the believer's commitments in light of its foundational story.

Hauerwas regards this type of "testing" (discipline) as a type of "natural law"[4] ([7], p. 120) perspective in that the community has settled upon certain convictions it believes are essential for sustaining its common life. Abortion (as a species of violence) is "unnatural" in the sense that it defies the "shape and nature of our lives appropriate to living in a world created by a gracious God" ([7], p. 120).

Such harsh language seems a type of absolutism, and lies at the heart of James Gustafson's concern with the "sectarian" quality of Hauerwas' writings ([2], pp. 83–95) – a charge Hauerwas has not answered very convincingly ([9], pp. 1–19). The problem is not simply a matter of opting for a tradition, admitting that it is one option among others or even offering this interpretation as the most appropriate and reasonable rendering of the biblical materials. Hauerwas' tone is that of the true believer; his moral interpretation seems not to be subject to inquiry or challenge.[5] It seems to assume that his viewpoint is the truthful rendering of the biblical story. The bias of a particular perspective is simply accepted and affirmed without apology or any discernible awareness of the possibility of being wrong at one or more points.

Whether Christian convictions are truthful or not can be tested by the faithfulness of the church ([9], p. 11), he says, which seems to mean that there can be no room for variety in response to practices such as abortion without removing oneself from the community. Sectarian rigidity expresses itself in a variety of ways.

The problem is not that Hauerwas is genuinely and openly Christocentric or confessional. It is that it does not betray the humility and openness that an emphasis on narrative would seem to require. The premise that Christian ethics is narrative-based is entirely supportable. I accept the premise that the Christ story is normative for Christians. But do such commitments of the mind and heart yield moral norms that are always valid for all believers under all circumstances? Are the norms so obvious as to be unmistakable and self-evident? Do those basic affirmations either lead to or require the sectarian spirit of such moral rigidity?

The contention that the operative image of the church is essentially sectarian (as defined above) for Hauerwas is enforced by the way in which he believes Christians are to learn family virtues. Prior to the question of abortion in Hauerwas' thought are attitudes toward childbearing, the role of children, and the relation of sex to procreation.

PERSPECTIVES ON CHILDBEARING

As Hauerwas sees it, the first question is "Who are we (in relation to sex and children) and who do we wish to be . . . and what kind of people must we be if we are to welcome children into the world?" ([7], p. 117). The answer relates both to character and to the community of which Christians are a part. The context for both the question and its answer is the active, redemptive work of God in Jesus Christ. God is working through community to redeem persons. That is the context in which Christians procreate. They understand that life is a sacred gift and that children are a symbol of hope.

Raising children expresses confidence that God "continues to welcome men and women into his kingdom" ([6], p. 133). Procreation in Christian marriage brings "into life new people whom God may also call to be a part of the ongoing story of the people of Jesus". Voluntary childlessness is a sign of faithlessness since community is a part of God's creation. Christians, he says, "have children because it is their duty – they are commanded to do so". Childbearing expresses the Christian willingness and "determination to exist as a people formed by the cross" ([4], p. 151), ([6], pp. 165, 226).

He goes so far as to argue that "we do not choose to have our children, nor do we have them because we cannot avoid them", which seems to imply opposition even to contraception. At issue for him is the survival of the Christian witness, since "we wish to continue and people those called into the world by a gracious God. . . . (This is the basis of our conviction about abortion, not that life is sacred, but that children should be regarded this way.)" ([4], p. 151).

The character of the community is therefore crucial for why and how we learn to rear our children. The Christian community is formed by the conviction that the power of this world is not the determining sway of our existence, but rather it is the power we find in the cross of Jesus Christ. Children witness to our determination to exist as a people formed by the cross even though the world wishes to deny that a people can exist without the power protected and acquired through the sword.

The problem of method is again apparent with Hauerwas' use of the story of Jesus as a correlative to or the narrative basis for his attitudes toward childbearing. What dominates his thought is a sectarian image of the church as pacific community, living in opposition to and from the world. The story of Jesus and his crucifixion is told in a highly selective or eccentric fashion that focuses primarily on non-violence and cross-bearing as the essential features of Jesus' life.

Moving from Jesus and his cross to childbearing is highly problematic, of course. Jesus was apparently unmarried and childless. His teachings relativized the importance of family and redefined the meaning of kinship. Brothers and sisters are those who see the kingdom and share its priorities, not those in blood relationships (Matt. 12: 48–50).

Jesus also spoke of those who "make themselves eunuchs for the sake of

the kingdom" (Matt. 19: 12) which seems to deny there is any "commandment" to bear children altogether. Jesus used children to identify the spirit
of trust and faith that are necessary to kingdom membership, but that had
nothing to do with procreation or the duty of childbearing. What Hauerwas
reflects, then, is not a careful delineation of the Gospel story by which he
does "narrative ethics" but the sectarian assumptions around which his image
of the church is fashioned.

CHRISTIAN SANCTIFICATION

The manner in which Hauerwas construes the doctrine of sanctification seems
further to enforce this (problematic) image of the church. Christians are
required "to be nothing less than a sanctified people of peace who can live
the life of the forgiven" ([7], p. 60). Sanctification is thus tied to transformation
which, in turn, is linked to particular and identifiable norms to which the
believer is to be conformed. The Christian is to adopt and adapt to the
practices of the community; the community is not there to develop people who
decide for themselves how to live. Living morally, he says, "is not a matter
of developing the ability to make well-reasoned decisions (which would make
it open to anyone) nor holding the right principles". It is a matter of being
initiated into a community that "shapes and forms life in a decisive and
distinctive way" ([9], p. 103). Christian development is a matter of learning
to "desire the right things rightly". That is not a matter of choice but of the
"slow training of our vision through learning . . . the story mediated to us
by masters who have learned what the story says. . . ." ([9], p. 103). When
facing difficult or problematic circumstances, the individual is able to "depend
on the judgment of others" – people of integrity who are "in a better position
to weigh matters sensibly . . .". ([4], p. 35).

Authority in moral matters is thus centered in the church, particularly in
the leaders (elders, priests, and pastors) of the community. The agenda with
liberalism causes Hauerwas to focus, not on the individual conscience shaped
by worship and the study of Scripture, but on the "wisdom" of the community, i.e. of the leaders. He does not deal with the seeming contradiction
between insisting that Christian moral thought includes highly subjective,
that is, personal commitments and his notion that emotional involvement
with a burdensome pregnancy makes it advisable to defer to the judgment of
others! Sanctification thus inculcates a posture of deference toward moral
authority in that the believer is to develop the ability to recognize the "saints"
and "masters" who have remained faithful to the story ([9], p. 103).

Hauerwas insists that sanctification is not a matter of being called to be
"morally good but rather to be faithful" ([9], p. 102), which seems to be a
distinction without a difference. He says it is not a matter of moral perfection but holiness ([7], p. 110). We are to have violence routed out of our
souls ([9], p. 104).

Hauerwas quite insightfully deals with the problem of self-deception and illusion on the part of those who rationalize their involvement in violence. This process often takes the form of "redescriptions which countenance coercion", which is as true of war as it is of abortion ([3], p. 46). He concedes that "an abortion might be a morally necessary, but sorrowful, occurrence". But the act must not be described as anything other than the killing of significant life. Instead of admitting the sin involved, he says, a self-deceiving policy is developed holding that the fetus, after all, is just another piece of flesh. He apparently prefers that the moral gravity of the action be admitted by calling it a justifiable killing ([4], p. 38).

That rationalization can and does take place we are all aware. Whether Hauerwas' polemics against the practice of abortion is clear of self-deception and illusion is less clear, however. He often uses the notion of "properly envisioning" in such a way as to imply that he has the inside track on clear thinking.

Hauerwas once criticized Barth's ethic of "command" for not making room for Christian growth ([6], pp. 3–4). The charge seems just as if not more fitting for Hauerwas. Making community norms the defining factors in faith is to adopt a notion of passivity in the believer's relation to the living presence of God. The church seems more important than the Holy Spirit in fashioning the life of the believer.

Saying, as Hauerwas does, that the community has priority over the individual is another way of saying that conscience has no liberty apart from the permission granted by the tradition. He understandably and rightly challenges the tenets of rugged individualism in ethics. He insightfully describes sin as "our active and willful attempt to overreach our powers . . . to live *sui generis* as if we are or can be authors of our own stories" ([7], p. 31). There are good reasons to reject the notion that ethical responsibility can be reduced to "each doing what seems good in one's own eyes", as he does.

The problem is that Hauerwas seems to substitute a type of moral imperialism as a corrective to what he perceives as the evils of individualism. Not only is his tone of writing paternalistic and authoritarian, the very structure of his ethics is rigid and unbending. For Hauerwas, the Christian life is not a pilgrimage of faith in which the believer is challenged to walk responsibly and develop skills of discernment and make decisions consistent with the "mind of Christ" and in the fear of the lord (Phil. 2: 5, 12b). His notion of sanctification has certain affinities to heteronomous ethics – moral action guides are imposed upon the believer by those having "superior" moral insight or authority.[6] The Church, he says, aims not at autonomy but at faithfulness ([9], p. 103).

In short, the Christian life is one of conformity to norms which are community-based. Being faithful is a matter of conforming to expectations and bending one's will to the stronger willed, or the masters and saints who serve as our source of moral wisdom and our guides through perplexing circumstances.

TRAGEDY AND THE CHRISTIAN LIFE

The final factor to be considered as a crucial variable in Hauerwas' attitude toward abortion is his notion of tragedy. Again, the central image deals with the story of Jesus, focusing on the crucifixion in such a way as to determine the way in which tragedy figures in the Christian moral life.

Hauerwas' first concern seems to be an effort to avoid the tragedy of wrong involvements. The four criteria for the (truthful) story we are to adopt aim at such deliverance. The story must have the power to: (1) release us from destructive alternatives; (2) provide ways of seeing through current distortions; (3) keep us from having to resort to violence; and (4) embody a sense for the tragic, i.e. how meaning might transcend power ([4], p. 35f.).

The story that does all this is that of Jesus "which enables us to speak of (God) and our relation to him truthfully", ([4], p. 79) and it is true to the very character of reality ([9], p. 102). Hauerwas does not attempt to prove the truth of Christ. He rightly posits that affirmation as central to being Christian and hardly subject to analysis or verification.

The one test of truth that is offered is that this "story is one that helps me to go on" ([4], p. 80) under adverse or even tragic circumstances. The Christian story provides strength to bear up and resist the temptation to resort to violence as a way to escape the burden and anguish of suffering. Hauerwas seems to construe the Cross of Christ as the symbol of human pain and tragedy. The story is understood to say that the faithful bear up under the burden rather than seek ways of escape. Trying to avoid the tragic only makes us more susceptible to it ([4], p. 73).

Several affirmations are made along the way to support that opinion and offer guidance to the faithful. For one thing, Christians are "a people who know who is in control" ([9], pp. 1–2). Just how Hauerwas believes God is "in control" is not clear. Surely he does not mean that God caused the tragedy, but that what seems tragic today (the cross) may serve God's redemptive purposes.

Further, he says, the cross instructs us in the meekness required if we are to be part of God's work. In the cross, "we see how the Kingdom will come into the world and we are charged to be nothing less than a cruciform people" ([9], p. 104). Thus, life may be a painful waiting in patience in a world of suffering and injustice.

Meanwhile, Christians are to reject the illusions and sentimentalities of the world. They have the story that "provides . . . skills to work us out of self-deceptions (the primary one of which is that we want to know the truth) and a vision to enable us to know what commitments we have made" ([4], p. 80). Suffering is, therefore, to be accepted by Christians and nobly, heroically borne. They can be strengthened by the image of Christ's bearing the cross. They are to do so without complaint, bitterness or resorting to violence, as Jesus did.

HAUERWAS ON ABORTION

Hauerwas' attitude toward abortion, therefore, is a reflection of prior assumptions regarding narrative and character, the nature of the church as moral community, the place of children and procreation, the nature of sanctification, and a Christian understanding of tragedy. For him, abortion is a practice which contradicts the Christian story, places one outside the Christian community, represents a failure of sanctification and is an attempt to overreach our powers and avoid the tragic. Since his comments on abortion reflect these assumptions, at least his approach to the issue can be better understood.

An instructive example is found in Hauerwas' discussion of the role of the pastor as prophet dealing with abortion. The story is that of a woman facing an unwanted pregnancy following a divorce. Though he knows of the woman's situation only through the pastor's telling, Hauerwas declares that she "decided to kill her child as a means of attacking her husband . . .". For him, her anger toward her husband took the form of destructive anger toward "the child" ([8], p. 46).

Not only was Hauerwas emphatic in his judgmentalism toward the woman, he condemns the young pastor for his unwillingness to confront her boldly with the moral demands of the church. The pastor, he said, "failed to help her interpret her situation in light of her relation to the Christian community and that community's support of her".

Numerous problems abound with Hauerwas' interpretation of this story, not the least of which is his pejorative attitude and unbending polemic about the woman's decision. To dismiss her action as a decision expressing anger at her husband is simplistic at best, but certainly inadequate. Claiming that she failed to understand the community's support of her is also to betray an idealism that is hardly in touch with the reality in most cases. The sad fact is that she may have been only in too much contact with the type of support tendered by the community! Good people are often long on verbal assurances of support but short on concrete actions that help with financial and personal burdens.

Even more problematic is the language employed to speak of her action. "She decided", he said, "to kill her baby". That he might believe that abortion is "killing babies" is certainly understandable on certain grounds. But to state it as if it were a self-evident fact, poses the problem of method in an acute way. No discussion of personhood or of how Christian theology yields the normative judgment that abortion is murder is given. He seems to believe that abortion is contemptible because he says it is. No further discussion seems warranted or necessary.

THE PROBLEMATICS OF HAUERWAS' ETHICS

Several problems can thus be discerned with Hauerwas' narrative method in ethical discourse. *One is his use of the polemic and diatribe in dealing with moral issues*. Strong references are made to words like "community", "story", and "Christ". These are positive reinforcements or appeals with which every believer can identify and would like to be associated. Negative enforcements are also strongly used: "killing", "killing babies", "anger", "neglect", "failure" (to appreciate the real support of the community), and "ignorance" (that is, not discerning the truth of the situation) and "faithlessness" (acting in ways contrary to what faith requires). All these are supposedly self-evident features of the story to which the Christian community is committed. One need not bother, therefore, with a careful delineation of points to be considered in reaching a moral judgment. The conclusion is foregone. Abortion is not an option.

Hauerwas' approach to Christian narrative ethics thus has no place for dialogue with deliberative ethics, which he rejects as the liberal (secular) influence of the Enlightenment. Secular approaches attempt to establish a framework based on reason and deliberation among people of good faith and sound minds to establish acceptable and/or appropriate options for public policy. The disinterested observer, identifiable principles, and common goals or the public good become important for moral reflection on issues dealing with procedures and/or public policy. Social and religious pluralism are also taken seriously as the context for ethical reflection. This type of "reasoning through" is anathema for Hauerwas. His concern for the distinctively Christian leaves no room for accommodating the secular perspective. They contradict what he sees as the Christian story that embodies identifiable practices that define both the nature of virtues and the kind of society (community) we are and should be. Open dialogue and mutual influence would be tantamount to sacrificing Christian distinctives.

What seems to the critic as a stubborn refusal to deal with the substantive issues at stake or show a reasoned basis for the position advocated so emphatically is rooted in this passionate concern for what is "Christian" as over against what is secular. The impression may well be left that the retreat into a claim about character and narrative seems little more than a pious dodge to escape the rigors of reasoned discourse.

The relation of narrative ethics as character/virtue-oriented and the deontological tone of Hauerwas' writings is a further concern. Hauerwas speaks the language of character shaped by story, but the narrow parameters he permits for a Christian's action pertaining to abortion is difficult to distinguish from the rhetoric of a pure rule deontologist.[7] In spite of his objections to principle-oriented thinking in ethical reasoning, are there not rather rigid norms that lie hidden in his construal of the Christian story? What others deal with in terms of rules and principles, Hauerwas seems to subsume under the rubric of practices and duty.

The problem remains, however, to give some account of how the story actually supports the particular norm or practice that is advocated so strongly. On this issue, the reader is given little specific guidance – perhaps because such an effort would tend in the direction of deliberative ethics. That there is some need for rules and principles for guidance in the Christian life seems both obvious and undeniable. They permeate the biblical story of the people of God and are vital for Christian guidance. The question is not whether there are rules and principles but how they are to be identified as to their nature and source, and how they are to be applied ([15], pp. 59–62).

A further problem is found in Hauerwas' construal of the nature of Christian sanctification. The image he provides is not one of growth toward "the mind of Christ" (Phil. 2: 5), but constant (faithful) conformity to the elements of character required by the church. This story teller is a moralist. A type of moral rigorism/absolutism pervades the pages of Hauerwas' writings, which seems contradictory to what one would expect of one espousing narrative ethics ([13], pp. 116, 135), or to what many see as the free obedience basic to New Testament ethics. The reason seems to lie in the fact that Hauerwas blends a Wesleyan concern for moral seriousness with generous doses of Mennonist communitarianism and Catholic authoritarianism.

The priority of community that establishes moral expectations seems also to replace the role of the Holy Spirit as a living, active presence in the believer's life. The authority of Scripture is also made secondary to that of the church ([10], p. 25). For him, "the Church, as the Body of Christ, stands first and is more full than Scripture". Giving priority to the moral authority of the Church helps to explain why he relies less on what the text may have to say on topics basic to his perspective than on those "practices" discernible in certain Christian traditions.

Hauerwas' concern for the distinctiveness of Christian ethics also leads him into the sectarian camp. He focuses on the observable action, the external deed, the forms of conduct. As he said, "the field of a story is actions (either deeds or dreams) or their opposite, suffering" ([4], p. 29). The claim that character and commitment predispose us not to do certain things seems both true and misleading. But to claim that abortion is morally forbidden as contrary to Christian character seems to betray a legalistic mindset, whether or not absolute and unbending rules are employed.

The Christian moral distinctive might more reasonably be construed in terms of grace and forgiveness rather than a harsh and unbending mandate. I fully agree with his point that "there are certain things that Christians just do not do". Among such actions are those that verbally abuse, humiliate, disempower, depersonalize, and inculcate false and/or inappropriate guilt in a woman who trusts us enough to seek our guidance in times of stress. To shame a woman into submission and obedience and thus to "coerce" her to bring a problem pregnancy to term seems one thing a Christian simply would not do.

Hauerwas' arguments concerning Christian marriage and child bearing also lack a biblical basis. How does one move from "the normative story of Jesus" to the notion that the church exists by procreation? Jesus was apparently single and childless. His teachings relativized the family and made it clear that genealogy had nothing to do with God's kingdom (Matt. 3: 9). By what authority then does Hauerwas claim that Christians are "commanded" to have children? The issue is especially acute where tragedy (rape, incest, fetal deformity, and maternal health) is involved. His position on abortion seems to substitute harshness in judgment for either discerning grace or compassionate insight. He seems to accept the notion of fetal personhood and its willful destruction as killing or murder as norms that are "givens" in the Christian community. No discussion of the norm or whether there is such a consensus ever appears in his writings. It is simply posited as an absolute and defining norm for faithful living within the Christian community.

Hauerwas' writings do yield clues as to the sources upon which these attitudes are based, however. *The Didache*, Early Church preachments, an image of the Christian community that draws upon the *koinonia* of Acts, and commitments to nonviolence most often associated with the historic peace churches are mentioned. Each of these sources pose critical problems for normative Christian ethics, of course, but he gives no reasons as to why *these* stories and not others are chosen. We are given emphatic judgments based upon many unexplicated assumptions.

One would at least expect greater care in dealing with methodological considerations if conclusions are going to be asserted with such stern certainty. By what authority might one claim that the fetus has the religious standing of a child? or that it is a sacred gift with an immunity from attack regardless of circumstance?

Simply to assert that high regard has been given to child-bearing in the Christian story hardly answers the substantive questions at stake. On what grounds does Hauerwas reach the conclusion that abortion is simply not an acceptable alternative for Christian women? Is it not entirely possible (probable?) that abortion has always been part of the Christian story, howbeit not one that is celebrated or advocated openly? The silence of the New Testament on this subject may well be eloquent testimony to the privacy in which such profound decisions are to be considered.

What account should be made of the woman's story or of her response to divine leadership? The issue is one of the nature of faith as response to the living God. Is faith only a type of obedience to the elderly wise of the Christian community? Or is it not more a matter of discerning God's will through the spiritual and community resources available to the believer in the context of perplexing and difficult life circumstances. Even if there were "community resources" available, is that the only factor that is to be weighed by the woman or that determines her decision? Surely financial support or child-care resources are not all that concerns the woman.

Certainly one can reach conclusions such as those espoused by Hauerwas by opting for certain stories within certain traditions. Very different conclusions are reached, however, by appealing to different stories and different traditions. How are the different claims or conclusions of the various Christian traditions to be reconciled? Even an appeal to "the biblical story" hardly yields unanimity of opinion or uniformity of action, as Hauerwas implies. At a minimum, one would think he is obliged to show how that story is as prohibitive of abortion as he assumes, especially since there is neither a clear prohibition nor a moral condemnation of the practice in Scripture.

A further question pertains to his understanding of tragedy, its place in the Christian story and its importance for Christian ethics. The moral meaning of tragedy is important but Hauerwas' use of the cross of Christ as a model for all or any human suffering is terribly problematic. How does he believe that "God is in control" when terrible or tragic things happen to Christians? Surely distinctions between various types of suffering and pain and their moral meanings are necessary. The vicarious sufferings of Mother Theresa and Dietrich Bonhoeffer are hardly of the same nature as the sufferings incurred by victims of rape or incest.

Furthermore, there is neither obvious virtue nor necessity in suffering that is not redemptive. Jesus' cross was an option; it was neither fate nor accident. He chose the cross. The question for Christians is how the cross figured in God's will for him – and how it might identify our suffering on behalf of the Kingdom. Some discernment is needed of the difference between the suffering that righteousness might require and that which has no redemptive purpose. Our "cross" is something to be chosen in being faithful to God's will, not something imposed by fate or blind circumstance.

A woman raped or a girl impregnated incestuously is hardly a crucifixion, but her suffering is significant morally and theologically. The "tragedy" was perhaps unavoidable. Certainly women are not morally culpable in becoming pregnant under such circumstances. Such a pregnancy compounds and complicates the violent assault she has experienced thus greatly extending and complicating her suffering. To her psychic, emotional and physical bruises is added the shocking awareness of a pregnancy that symbolizes hate and injury, not love and commitment. She has been "thrown" into a major life crisis. Her life is uniquely at stake – whether or not physical complications threaten her with death. Her future directions and lifeplan may be significantly altered if the pregnancy comes to term. Pregnancy is not just something that happens to a woman and goes away; it is a life-changing event to be "mother", not just woman.

It is little comfort to remind a woman faced with a burdensome and threatening pregnancy of the story of Jesus who bore his cross and did not resort to violence against the Jews or Romans. Her despair and desperation may be deepened when she is told that she has no alternative but to suffer heroically. To that paternalistic and authoritarian insult is often added the religious threat that to do otherwise is to place herself outside the Christian story. Playing on her misgivings and creating guilt for considering abortion, compounds the tragedy even more.

To be sure, the woman may retain the pregnancy. But it should neither be a matter of necessity nor a matter that all other options are morally closed. Her pregnancy may be for her a moment before God which is to be accepted courageously. She may even do so vicariously – for the sake of the child to be, or for the sake of childless couples who want to adopt, or simply as a shift in her own lifeplan. For this suffering there is sustaining grace.

But there is also delivering grace for those who perceive the divine will for their lives in different terms. These women are given permission to act against evil and the perpetuation of the violence and injustice she has already experienced. Hauerwas comes close to identifying fatalism with the divine will. Stoicism is not Christian however much it might comfort the suffering. Believers are to interact with history, shaping it in the light of the future willed by God. They are not simply to accept the tragedy of problem pregnancy as if it were the sign of the cross.

Another problem is that Hauerwas' construal of the Christian story regarding abortion is both anthropocentric and androcentric. His images and norms are drawn from male-dominated traditions. Stories of women and the agonies associated with pregnancy and child bearing are needed to enlarge his vision and inform his ethics. He is right about the need for a truthful story to enable us "to see through current distortions" ([4], p. 35), but he shows little awareness of the misogyny that often attends the church's story of Jesus and its humiliating effect on women. He ends by identifying faith with keeping fidelity to that distorted story which is a counsel to women to remain pregnant, submissive and subordinate.

But the story of faith by a woman facing a burdensome pregnancy and crying out for relief is vital to hear. Her redemption is not in childbearing [14].[8] It may be in the form of a caring physician to deliver her from an odious and threatening pregnancy. She may rejoice in that liberation as a sign of the redemptive will of God. That may truly be a moment of grace and delivering love. It may be a sign of resurrection, which is, after all, the Christian word beyond the cross.

CONCLUSION

This exploration/examination of Hauerwas' approach to abortion has isolated several features basic to his understanding of narrative ethics and to his conclusions regarding the moral acceptability of abortion. The fact that the assumptions basic to his argument represent highly sectarian perspectives indicates why his opinions are so highly suspect in many circles. The fact that secular bioethicists find them idiosyncratic is not at all surprising; many Christian ethicists find them intriguing, but hardly persuasive.

That Christian ethicists see the issues in different ways sharply illustrates the need for dialogue, not dogmatism on critical and divisive issues like abortion. Any pronouncement regarding an issue in ethics needs the humility of recognizing the limited, biased or relative point of view in every human perspective. The "data" used by theologians are terribly diverse. Moreover, they disagree among themselves as strongly as they disagree with those who argue on secular grounds. Christians certainly have no monopoly on truth, as the sordid story of the Church's action toward Galileo so vividly illustrates.

The problem with Hauerwas' approach, as I see it, is not with narrative as an approach to Christian ethics, but with his particular construal of narrative ethics and what he calls "the Christian story". He invests "the Christian story" with meanings and images I find alien to the New Testament portrait of Jesus and foreign to the nature of the Christian life. Just how he moves between the narrative(s) of Scripture and his understanding of the narrative structure of the Christian life is an area he leaves unexplained. Adopting an image of the church that is highly sectarian and rigidly moralistic and authoritarian (again without explaining why or showing how that "story" is to be preferred above all others) is also a major problem.

Further, Hauerwas' style of arguing from a type of "special knowledge" by which no arguments against his can be launched is especially problematic. Surely that is why "biblical authority", that is, specifically struggling with and interpreting the biblical material, is necessary and indispensable for Christian ethics. In spite of the difficulties of relating the Bible to Christian ethics, it has the great advantage of allowing a common point of reference for discussing *the story* so basic to being Christian. All members of the Christian community may reflect upon the nature of God's grace by which we live, the divine will to which we are responsible, and the church as a moral community by which we are shaped, given direction, support and encouragement.

Finally, the narrative structure of the Christian life requires closer attention to the rich reportoire of stories that define the story of the people of God. It creates a community in which grace is demanding, delivering *and* permitting. Those women who opt for abortion are or should be surrounded by people who understand tragic and exceptional circumstances,and who provide insight and wisdom as to the presence and power of grace. That community is much richer in its variety of perspective and more generous in its grace and forgiveness than that envisioned by Hauerwas, it seems to me. Choosing an abortion is not the same as rejecting the Christian story, nor is it to opt out of the community created by the death and resurrection of Jesus. It may well be the result of understanding those central and defining events in even more profound and personal ways.

University of Louisville
Louisville, Kentucky, U.S.A.

NOTES

1. Hauerwas regards it a fundamental mistake to approach Christian ethics as a calculation of the right thing to do. He says this approach is inspired by the scientific ideal of objectivity and universalizability. He thus rejects such approaches as the categorical imperative, the ideal observer, and the original position.
2. Hauerwas admits that "there may well be circumstances when abortions are morally permissible . . ." but he gives no clues for knowing when that might be the case. More typically he regards abortion as "morally tragic", and speaks of "the Christian prohibition of abortion".

He says that "the Christian way of life forms people in a manner that makes abortion unthinkable".

3. There is nothing about a narrative method as such that leads to a strongly prohibitive approach to abortion. The notion that Christian character and moral actions are shaped by the stories to which we belong I believe to be fundamentally sound. The issue is the manner in which "the story of Jesus" or "the Christian story" is to be construed.

4. Hauerwas rejects traditional notions of natural law theory but retains the terminology by investing it with new meaning. For him, "'Natural law' really names those moral convictions that have been tested by the experiences of the Christian community and have been judged essential for sustaining its common life" ([7], p. 120).

5. Interestingly, he says that theological convictions are open to challenge since they involve truth claims. That underscores the value of narrative instead of doctrines. See ([9], p. 9).

6. To argue that our moral identities are formed by the communities in which we participate hardly removes the objection concerning heteronomy. Saying that persons are "willing" participants in authoritarian communities is also problematic as numerous studies of power and submission have shown. Abortion is a special case in point since the moral norm Hauerwas espouses is typically enforced by "leaders" who are "external" judges. They are *men* and thus will never know what it means to be pregnant or experience the threat and burden of a problem pregnancy. These "leaders" are thus deprived of a perspective based in experience – another dimension of a truthful narrative.

7. Hauerwas is neither a pure rule deontologist nor a formal rule deontologist; his deontology is that of the duty or obligation to adopt the practices he believes define the Christian community.

8. Aida Spencer's commentary on 1 Timothy 2:15 points out that the genitive following *dia* should be translated "through the childbearing". Were it in the accusative, it would be "because of". Timothy's allusion is apparently to Mary's bearing the child Jesus, who saves us. Women are saved by faith in Christ, not by having babies, which would be a crass form of works righteousness.

BIBLIOGRAPHY

1. Gustafson, J.: 1971, *Christian Ethics and the Community*, Pilgrim Press, Philadelphia.
2. Gustafson, J.; 1985, 'The Sectarian Temptation: Reflections on Theology, the Church and the University', *Proceedings of the Catholic Theological Society* 40, 83–95.
3. Hauerwas, S.: 1974, *Vision and Virtue*, Fides Press, Notre Dame.
4. Hauerwas, S.: 1977, *Truthfulness and Tragedy*, University of Notre Dame, Notre Dame.
5. Hauerwas, S.: 1980, 'The Church in a Divided World', *Journal of Religious Ethics* 8(1).
6. Hauerwas, S.: 1981, *A Community of Character*, University of Notre Dame, Notre Dame.
7. Hauerwas, S.: 1983, *The Peaceable Kingdom*, University of Notre Dame, Notre Dame.
8. Hauerwas, S.: 1985, 'The Pastor as Prophet: Ethical Reflections on an Improbable Mission', in E. Shelp and R. Sunderland (eds.), *The Pastor as Prophet*, Pilgrim Press, Philadelphia.
9. Hauerwas, S.: 1988, *Christian Existence Today*, Labyrinth Press, Durham.
10. Hauerwas, S.: 1993, *Unleashing the Scripture*, Abingdon, Nashville.
11. Hauerwas, S.: 1994, 'Why I am Neither a Communitarian nor a Medical Ethicist', *Hastings Center Report*, Special Supplement.
12. McClendon, J.: 1986, *Systematic Theology: Ethics*, Abingdon, Nashville.
13. Nelson, P.: 1987, *Narrative and Morality: a Theological Inquiry*, Penn State Press, University Park.
14. Spencer, A.B.: 1985, *Beyond the Curse: Women Called to Ministry*, Thomas Nelson Publishers, Nashville.
15. Stassen, Glen H.: 1980, 'Critical Variables in Christian Social Ethics', in P. Simmons (ed. and cont.), *Issues in Christian Ethics*, Broadman, Nashville, pp. 57–76.

THOMAS A. SHANNON

Genetics and Freedom:
A Critique of Sociobiological Claims

I. INTRODUCTION

GOD'S GRANDEUR
Gerard Manley Hopkins

The world is charged with the grandeur of God.
   It will flame out, like shining from shook foil;
   It gathers to a greatness, like the ooze of oil
Crushed. Why do men then now not reck his rod?
Generations have trod, have trod, have trod;
And all is seared with trade; bleared, smeared with toil;
And wears man's smudge and shares man's smell: the soil
Is bare now, nor can foot feel, being shod.

And for all this, nature is never spent;
There lives the dearest freshness deep down things;
And though the last lights off the black West went
Oh, morning, at the brown brink eastward, springs
Because the Holy Ghost over the bent
World broods with warm breast and with ah! bright wings.
   ([11], p. 27).

The question of whether matter and its structure is a sufficient explanation of human behavior is a recurring question within human history.[1] Within the last decades this question has received a new emphasis and focus from the discipline of sociobiology which approaches the question from a biological perspective. But while the focus has shifted from physics to biology as the normative framework, the answer remains relatively constant: one does not need to look beyond matter and its structure to explain human behavior in all its complexity and subtility.

To begin on issues in genetics and sociobiology with a poem by Gerard Manly Hopkins rather than *In Memoriam* by Tennyson with its perhaps more relevant line "Nature, red in tooth and claw" reveals the perspective I will

E. E. Shelp (ed.), Secular Bioethics in Theological Perspective, 177–202.
© *1996 Kluwer Academic Publishers. Printed in the Netherlands.*

use to critique various claims from sociobiology. This perspective rejects a materialistic perspective and proposes instead a philosophical/theological understanding of the reality of matter and human experience. Thus, I will argue against materialistic interpretations of human experience made by leading sociobiologists.

The above poem by Gerard Manly Hopkins suggests a twofold argument. First, the line "there lives the dearest freshness deep down things" speaks particularly to sociobiological discussions of evolution and genetics, especially with respect to an understanding of matter and the location of humans within matter. In particular the poem points to experiences of transcendence within the material world that challenge the materialist framework. Second, the thought of Hopkins was deeply inspired by the philosopy of John Duns Scotus, the Franciscan philosopher of the late Middle Ages. The angle of vision from which I examine and argue against the claims of sociobiology will be that of Scotus, whose analysis of human experience and freedom has, in my judgment, considerable insights to bring to issues raised by sociobiology. There are two reasons for this. First, some of the issues he discusses are particularly relevant to contemporary discussions of human freedom and agency. Second, Scotus discusses issues that are quite similar to – if not actually identical with – issues in sociobiology. While I clearly recognize that his world is not our world, nonetheless, Scotus brings a perspective that is helpful in evaluating the claims of sociobiology.

In particular, then, I present a Scotistic account of transcendence and freedom within human experience that challanges the reductionistic and mechanistic claims made by E.O. Wilson and Richard Dawkins. This essay will speak to several claims made by them from a philosophical/theological perspective. The essay itself is neither a review of sociobiology, in general, nor an attempt to summarize concepts from the wide ranging literature in sociobiology. Rather my strategy is to examine writings by two significant figures in the field of sociobiology – Wilson and Dawkins – who have made particular efforts to present their ideas to as broad an audience as possible. Because of their status and because the ideas of these men are easily accessible in the public domain, I use their writings as the basis for my analysis.

Two themes, then, set the tone for this essay. First, the line "there lives the dearest freshness deep down things" speaks particularly to sociobiological discussions of evolution and genetics, especially with respect to an understanding of matter and the location of humans within matter. Thus the discussion of materialism will be a major feature of the essay. Second, in the penultimate line, Hopkins suggests the reason for this freshness: the brooding of the Holy Ghost over the bent world. I wish to highlight the word "bent". This word is interesting from two perspectives. First "bent" is reminiscent of the Augustinian/Bonaventurian understanding of the consequence of original sin: humans are curved in upon themselves. Second, one can think of "bent" from sociobiology's assertion of the selfish gene: human action focused on its own good. A question, then, is whether and how this bent world might be straightened, particularly in view of sociobiology's claim of genetic selfishness.

My task is to evaluate the adequacy of several claims of sociobiology. I shall challenge, from a Scotistic perspective, the adequacy of scientific materialism as an explanatory framework for human behavior, the concept of altruism as presented by sociobiology, and, finally, sociobiological perspectives on human freedom.

II. Scientific Materialism or a "Dearest Freshness Deep down Things"?

In this section I explore sociobiology's, and particularly Wilson's, claims about scientific materialism. In particular I examine the adequacy of scientific materialism in capturing the sufficiency of matter as described by Wilson as the ground for evolutionary development.

A. *Scientific Materialism in Sociobiology*

Although this concept appears late in Wilson's *On Human Nature*, it is a key principle which provides the overarching framework for many of the ideas in sociobiology, "the systematic study of the biological basis of all social behavior" ([19], p. 4). Scientific materialism is "the view that all phenomena in the universe, including the human mind, have a material basis, are subject to the same physical laws, and can be most deeply understood by scientific analysis". ([18]. p. 221). The core of scientific materialism is the evolutionary epic whose minimum claims are:

> that the laws of the physical sciences are consistent with those of the biological and social sciences and can be linked in chains of causal explanation; that life and mind have a physical basis; that the world as we know it has evolved from earlier worlds obedient to the same laws; and that the visible universe today is everywhere subject to these materialist explanations ([18], p. 201).

Scientific materialism is a mythology and "the evolutionary epic is probably the best myth we will ever have" ([18], p. 201) and it can be "adjusted until it comes as close to truth as the human mind is constructed to judge the truth" ([18], p. 201).

Of critical importance is a discussion of matter, the ultimate grounding – so to speak – of evolution. In Wilson's theory, matter is all that is and all that is needed to account for all activity – insect or animal, private or social. For Wilson, matter is most creatively expressed in the gene, the basic unit of heredity and "a portion of the giant DNA molecule that affects the development of any trait at the most elementary biochemical level" ([18], p. 216). Thus we need to examine human nature through biology and the social sciences. This will lead us to an understanding of the mind "as an epi-phenomenon of the neuronal machinery of the brain. That machinery is in turn the product of genetic evolution by natural selection acting on human populations for hundreds of thousands of years in their ancient environments" ([18], p. 195).

B. *The Transcendent Potential of Matter*

But is matter only matter, inert particles interacting according to the laws of physics and/or chemistry or is there another level?

One theory explaining the interaction of particles of matter such as elec-trons and positrons is hylosystemism which holds that "all bodies, or at least non-living bodies, are composed of elementary particles or hylons which are united to form a dynamic system or functional unit" ([20], p. 98). In this context, system refers to "a functional nature, possessing new powers" ([20], p. 105). When put into various combinations or when actualized under various conditions, these elementary particles form new systems educed from the matter and the properties of this new system:

> are not simply the arithmetical sum of the actual properties manifested by these hylons in isolation for the property of any given system such as the nucleus or the hydrogen atom . . . is rooted proximately in the new powers of the respective system, powers which, though ultimately reducible to the two or more hylons that function as essentially ordered causes, exist only virtually in the individual hylons ([20], p. 105).

Consequently, the properties of individual particles seen in isolation can never tell us the full range of these particles when combined into a system. Therefore, within matter lies a range of possibilities that emerge or are actualized only when these particles are put into a system or when a previous system is restructured.

What are the implications of such a theory? One is presented by Karl Rahner who argues that we are the beings "in whom the basic tendency of matter to find itself in the spirit by self-transcendence arrives at the point where it definitely breaks through" ([15], p. 160). For Rahner this means that "matter develops out of its inner being in the direction of spirit" ([15], p. 164). This becoming, a becoming more rather than becoming other, must be "effected by what was there before and, on the other hand, must be the inner increase of being proper to the previously existing reality" ([15], p. 164). This notion of becoming more is a genuine self-transcendence, a "transcendence into what is substantially new, i.e., the leap to a higher *nature*" ([15], p. 165).

While Rahner does not argue that life, consciousness, matter and spirit are identical, he does argue that such differences do not exclude develop-ment:

> In so far as the self-transcendence always remains present in the par-ticular goal of its self-transcendence, and in so far as the higher order always

embraces the lower as contained in it, it is clear that the lower always precedes the actual event of self-transcendence and prepares the way for it by the development of its own reality and order; it is clear that the lower always moves slowly towards the boundary line in its history which it then crosses in actual self-transcendence ([15], p. 167).

For Rahner, then, the human is the "self-transcendence of living matter" ([15], p. 168). On the one hand, Rahner describes this as the cosmos becoming conscious of itself in the human. Yet on the other hand again, this self-transcendence of the cosmos reaches:

> its final consummation only when the cosmos in the spiritual creature, its goal and its height, is not merely something set apart from its foundation – something created – by something which receives the ultimate self-communication of its ultimate ground itself, in that moment when this direct self-communication of God is given to the spiritual creature in what we – looking at the historical pattern of this self-communication – call grace and glory ([15], p. 171).

Another presentation of matter as the ground of new potentialities is presented by Lindon Eaves and Lora Gross who argue for a dynamic and holistic conception of matter that emphasizes the "unity of matter, life, and energy and understands nature as a profoundly complex, evolving system of intricately interdependent elements" ([7], p. 226). They argue for an understanding of vitality in matter which gives it depth and intensity, value, and the inclination toward organization.

Eaves and Gross operate from a biological and specifically genetic perspective that "seeks a new framework for its comprehension that does justice to all the so-called higher aspects of human consciousness in a phylogenetic and ontogenetic framework" ([7], p. 274). This perspective focuses on the mechanisms of inheritance which "have within themselves the probability of presenting new transcendent possibilities for action within history" ([7], p. 278). Thus they argue that surprise is inherent in nature and develop a view of nature itself as gracious and also argue, similarly to Rahner, that "genetics provides *a basis for grace within the structure of life itself*" ([7], p. 274).

This position serves as the basis for a rejection of crude determinism for:

> the material processes of life have produced a person who transcends all conventional definitions of personhood to the point where the term *freedom* is the best we have available ([7], p. 275).

This gives rise to two consequences: first, "culture creates conditions for completion in community that would otherwise be impossible in a mere aggregation of individuals" ([7], p. 277); second, "recognition that the conditions of life are such that the process that produces pain, in the sense of genetic disease, is also the process that maintains life in the cosmos" ([7], p. 278).

This second point is critical in that it highlights the value of genetic diversity and provides the ground for criticizing simplistic models of genetic waste, unfitness and disease. A similar point emerges from a consideration of the multiplicity of forms and species.

> There are many forms which do not constitute a value or an advantage in the struggle of life; they are useless in this sense, and for that reason they are beautiful. Beauty is a factor that is not necessitated by lower needs, but is something that supposes the liberty of artistic creation ([2], p. 157).

This view of matter and evolution is in the tradition of Augustine and his follower Bonaventure who saw history as a most beautiful song, a *"pulcherrimum carmen* which has been played by the divine Wisdom since the first organisms were called into existence, and of which our present forms are but one scene" ([2], p. 157). Or, as the *Book of Proverbs* says of Wisdom:

> I was by his side, a master craftsman, delighting him day after day, ever at play in his presence, at play everywhere in his world . . . (8:30).

III. THE SELFISH GENE OR THE GOOD OF OTHERS?

A. *Genetic Selfishness*

One of the most widely known and quoted claims of sociobiology is that of genetic selfishness. This claim flows from the evolutionary strategy of reproductive success in which genes seek to replicate themselves. Thus the individual is but DNA's way of making more DNA. This view was promulgated most clearly in Dawkins' book *The Selfish Gene*, the main argument of which was that "a predominant quality to be expected in a successful gene is ruthless selfishness. . . . Much as we might wish to believe otherwise, universal love and the welfare of the species as a whole are concepts which simply do not make evolutionary success" ([6], pp. 2–3).

Wilson understands altruism to be the central theoretical problem of sociobiology. This is so because in a "Darwinist sense the organism does not live for itself. Its primary function is not even to reproduce other organisms; it reproduces genes, and it serves as their temporary carrier" ([19], p. 3). This occurs through natural selection "a process whereby certain genes gain representation in the following generations superior to that of other genes located at the same chromosome positions" ([19], p. 3). Thus the organism is but DNA's way of making more DNA, and the individual but the vehicle of the genes.

In this context, the question is how can altruism – "self-destructive behavior performed for the benefit of others" ([18], p. 213) – possibly evolve through natural selection. Obviously this behavior reduces personal fitness and would seem to lead to the loss of the gene or genes responsible for that behavior. Wilson finds the answer to this question in kinship:

> if the genes causing the altruism are shared by two organisms because of

common descent, and if the altruistic act by one organism increases the joint contribution of these genes to the next generation, the propensity to altruism will spread through the gene pool. This occurs even though the altruist makes less of a solitary contribution to the gene pool as the price of its altruistic act ([19], p. 3).

Wilson seeks to show that "the impulse need not be ruled divine or otherwise transcendental, and we are justified in seeking a more convenient biological explanation" ([18], p. 152). Though Wilson notes that specific forms of altruism are culturally determined, he argues that the sociobiological hypothesis "can explain why human beings differ from other mammals and why, in one narrow aspect, they more closely resemble social insects" ([18], p. 153).

Wilson further distinguishes two forms of cooperative behavior. First is what he terms hard-core altruism: "the altruistic impulse can be irrational and unilaterally directed at others; the bestower expresses no desire for equal return and performs no unconscious actions leading to the same end" ([18], p. 155). Here the responses are unaffected by social reward and punishment and tend to serve the "altruist's closest relatives and to decline steeply in frequency and intensity as relations become more distant" ([18], p. 155).

Second is soft-core altruism: the altruist "expects reciprocation from society for himself or his closest relatives. His good behavior is calculating . . ." ([18], pp. 155–156). Thus soft-core altruism is essentially selfish and influenced by cultural evolution. The psychological vehicles for this behavior are "lying, pretense, and deceit, including self-deceit, because the actor is most convincing who believes that his performance is real" ([18], p. 156).

In Wilson's perspective, soft-core altruism is the key to human society because it broke the constraints on the social contract imposed by kin selection. Reciprocity is the key to the formation of society. Hard-core altruism, on the other hand, is the "enemy of civilization" ([18], p. 157). This favors kin selection, the favoring of one's own relatives, and permits only limited global cooperation. Thus he says,

> Our societies are based on the mammalian plan: the individual strives for personal reproductive success foremost and that of his immediate kin secondarily; further grudging cooperation represents a compromise struck in order to enjoy the benefits of group membership ([18], p. 199).

This gives Wilson a basis for optimism, for he thinks humans are:

> sufficiently selfish and calculating to be capable of indefinitely greater harmony and social homeostasis. This statement is not self-contradictory. True selfishness, if obedient to the other constraints of mammalian biology, is the key to a more nearly perfect social contract ([18], p. 157).

These other constraints are learning rules and emotional safeguards. Thus, honor and loyalty are reinforced while cheating, betrayal, and denial are universally rejected. Thus it seems that learning rules, based on innate, primary reinforcement, led human beings to acquire these values and not others with reference to members of their own group. . . . I will go further to speculate that the deep structure of altruistic behavior, based on learning rules and emotional safeguards, is rigid and universal. It generates a set of predictable group responses. . . . ([18], pp. 162–163).

Thus soft-core altruism provides the basis for various social allegiances, shifting though they may be. The critical distinction is the ingroup and the outgroup, the line between which fluctuates continually. But this is our social salvation for if hard-core altruism were the basis of social relations, our fate would be a continuous "intrigue of nepotism and racism, and the future bleak beyond endurance" ([18], p. 164). Soft-core altruism provides an optimistic cynicism which can give us the basis of a social contract. Such behavior has been "genetically assimilated and is now part of the automatically guided process of mental development" ([18], p. 167). Thus genes hold culture on a leash and though the leash is long, "inevitably values will be constrained in accordance with their effects on the human gene pool" ([18], p. 167).

Important in Dawkins' perspective is the affirmation that the focus is behavior not motive: the effects of one's act, not one's subjective dispositions. Thus, in the definition of altruism as behavior "to increase another such entity's welfare at the expense of its own" ([6], p. 61), welfare is understood as one's chance of survival. One looks at outcome, not motives. Thus, a selfish gene tries "to get more numerous in the gene pool. Basically the gene does this by helping to program the bodies in which it finds itself to survive and to reproduce" ([6], p. 95). However – and this is a key issue for this section – "a gene might be able to assist *replicas* of itself which are sitting in other bodies. If so, this would appear as an act of individual altruism but it would be brought about by gene selfishness" ([6], p. 95).

The key way in which such genetically altruistic acts occur is through kin selection or within-family altruism. A genetically altruistic act is one which increases the greatest net benefit to one's genes, i.e., insures the highest success rate for a particular gene. As Dawkins phrases it:

> A gene for suicidally saving five cousins would not become more numerous in the population, but a gene for saving five brothers or ten first cousins would. The minimum requirement for a suicidal altruistic gene to be successful is that it should save more than two siblings (or children or parents), or more than four half-siblings (or uncles, aunts, nephews, nieces, grandparents, grandchildren), or more than eight first cousins, etc. Such a gene, on average, tends to live on in the bodies of enough individuals saved by the altruist to compensate for the death of the altruist itself ([6], p. 100).

Thus Dawkins concludes: "I have made the simplifying assumption that the individual animal works out what is best for his genes" ([6], p. 105). This is essentially what Wilson calls hard core altruism and it is important to recall that such behavior is "the enemy of civilization" ([18], p. 157). Soft-core altruism, recall, is what makes society possible, though to a limited degree only. Thus for Wilson, the:

> most elaborate forms of social organization, despite their outward appearance, serve ultimately as the vehicles of individual welfare. Human altruism appears to be substantially hard-core when directed at closest relatives, although still to a much lesser degree than in the case of the social insects and the colonial invertebrates. The remainder of our altruism is essentially soft. The predicted result is a melange of ambivalence, deceit, and guilt that continuously troubles the individual mind ([18], p. 159).

This perspective seems to leave us in a rather melancholy state at best and total despair at worst. From a biological perspective, both Wilson and Dawkins seem to have placed us squarely in the middle of a Hobbesian world. Indeed as Dawkins says, "I think 'nature red in tooth and claw' sums up our modern understanding of natural selection admirably" ([6], p. 2).

B. *Human Altruism*

I argue that sociobiologists have made a major mistake in their use of the term altruism. My issue is the term, not the behavior – although my concern is not exclusively semantic. That is, while the behaviors described are biologically accurate – in so far as they stick to biology – the significance of these behaviors also has been misinterpreted primarily because of the sociobiologists' almost ideosyncratic use of the term altruism. And it is because of this that they have gotten themselves into their Hobbesian world.

This is where the medieval philosopher John Duns Scotus and his distinction between the *affectio commodi* and *affectio justitiae* can be helpful. Since Scotus died in 1302 it is obvious that he had no knowledge of the theory of evolution nor a concept of what sociobiologists would refer to as a reproductive strategy. Thus I am not attempting to bootleg any such theories into his thought. Nor am I attempting to use his ideas as a Procrustian bed with which to shape contemporary ideas. Rather, my sense is that Scotus has some insights that can help clarify the conundrum into which the sociobiologists seems to have gotten themselves.

Duns Scotus begins with two distinctions. First is the concept of a nature: a principle of activity by which an entity acts out or actualizes its reality. A being's nature is the reason why an entity acts as it does. Or as he says: "the potency of itself is determined to act, so that so far as itself is concerned, it cannot fail to act when not impeded from without" (*Quaestiones in Metaphysicam* I, q. 15, A. 2 in [21], p. 151). A nature is essentially the reason why an entity acts as it does.

A will, on the other hand, "is not of itself so determined, but can perform either this act or its opposite, or can either act or not act at all" (*Quaestiones in Metaphysicam* I, q. 15, a. 15 in [21], p. 151). Thus, the reason why this act was done as opposed to another is that the will is the will and can elicit an act in opposite ways. Following Anselm, Scotus distinguishes two movements in the will as the *affectio commodi* – the inclination to seek what is advantageous or good for one's self – and the *affectio justitiae* – the inclination to seek the good in itself.

In this section I focus on the *affectio commodi*, the will to do what is to our advantage, perfection or welfare. This affection or inclination is a nature seeking its own fulfillment. For Scotus, this *affectio commodi* is not an elicited act. Rather it is a natural appetite necessarily seeking its own perfection. As Scotus says:

> That it does so *necessarily* is obvious, because a nature could not remain a nature without being inclined to its own perfection. Take away this inclination and you destroy the nature. But this natural appetite is nothing other than an inclination of this sort to its proper perfection; therefore the will as nature necessarily wills its perfection, which consists above all in happiness, and it desires such by its natural appetite (*Ordinatio* IV, suppl. dist, 49, qq. 9–10 in [21], p. 185).

Allan B. Wolter provides an interesting commentary on this concept:

> we could say all striving, all activity stems from an imperfection in the agent. As the etymological derivation of the word itself suggests, nature [from 'nascor', to be born] is literally what a thing was born to be, or more precisely, born to become, for nature as an active agent is essentially dynamic in a Faustian sense. It is restless until it achieves self-perfection. Since what perfects a thing is its good and since this striving for what is good is a form of love, we could say with Socrates that all activity is sparked by love ([22], p. 150).

This love, however, is neither objective nor directed to the good of another, regardless of whether this other being might be a kin. It is self-centered and directed to seeking its own welfare. As Wolter further comments,

> If at times we encounter what seems to be altruistic behavior in the animal world, for instance, it is always a case where the "nature" or "species" is favored at the expense of the individual. But nature, either in its individual concretization or as a self-perpetuating species, must of necessity seek its own perfection. Such is its supreme value and the ultimate goal of all its loves ([22], p. 150).

As Wolter interprets Scotus here, the issue is that an individual entity in its actions seeks its own good or what is to its advantage. We should not be surprised at this because this is what a nature does, whether looked at as an individual representative of the species or as the species as a whole. The *affectio commodi* drives the being "to seek his perfection and happiness in all he does" ([22], p. 151).

What is significant about this perspective – particularly in the context of the sociobiologists – is that for Scotus – indeed for the entire classical philosophical tradition from Plato forward – this inclination to seek one's own perfection is a *good*. It is "not some evil to be eradicated. For it too represents a God-given drive implanted in man's rational nature which leads him to seek his true happiness . . ." ([22], p. 151). In fact, to ignore our perfection or to give it no standing in our actions is an act of injustice to ourselves.

What Wilson and Dawkins refer to as altruism understood in the context of genetic selfishness is what Scotus refers to as the *affectio commodi*. The significance of the Scotistic position is, on the one hand, that he too sees the same kind of tendency present in human nature as do the sociobiologists but, on the other hand, he, together with the entire philosophical tradition to that time, sees that behavior as a good because it achieves the perfection of the individual and the species. That is, for Scotus, the *affection commodi* is a constitutive part of human nature that leads us to seek our completion or perfection as a human. Precisely because this affection leads to our perfection it is a good.

There is, however, a critical difference between Scotus and the sociobiologists. For the sociobiologists, the behavior comes from evolutionary success whereas for Scotus the cause is the will of God expressed in creation. Nonetheless, while the origin is different, the behavior is the same. Part of the difference surely lies in both philosophical and theological frameworks. But another part of the difference lies in that Scotus sees self-perfecting behavior as a good while the sociobiologists describe this as selfish – which of course has a highly negative connotation.

But there remains this issue raised by the sociobiologists: is such genetically selfish activity the only possible mode of human activity? Or as Scotus would phrase it, can we see a good beyond ourselves and our perfection?

IV. DETERMINED TO ACT OR FREE TO ACT?

A. *Determinism*

The context of scientific materialism reasserts itself in Wilson's definition of free will:

> To the extent that the future of objects can be foretold by an intelligence which itself has a material basis, they are determined – but only within the conceptual world of the observing intelligence. And insofar as they can make decisions of their own accord – whether or not they are determined – they possess free will ([18], p. 171).

Wilson uses the example of a bee. If we were to know the properties of small objects, the bee's nervous system, its behavioral characteristics, and its personal history and could put this on a computer program, we could predict the bee's flight.

> To the circle of human observers watching the computer read-out, the future of the bee is determined to some extent. But in her own "mind" the bee, who is isolated permanently from such human knowledge, will always have free will ([18], p. 73).

The same is true for humans, insofar as their behavior can be specified. However, because of the complexity of human behavior and technical limitations, and perhaps, the capacity of intelligence in general, such specification and prediction of human behavior is practically impossible. Wilson concludes:

> Thus because of mathematical indeterminacy and the uncertainty principle, it may be a law of nature that no nervous system is capable of acquiring enough knowledge to significantly predict the future of any other intelligent system in detail. Nor can intelligent minds gain enough self-knowledge to know their own future, capture fate, and in this sense eliminate free will ([18], pp. 73–74).

For Wilson free will is either indeterminacy or unpredictability and is a function of a technical inability either to know all the variables or – should they be known – to program them in a meaningful way.

B. *Freedom*

Responding to Wilson brings us to the other critical distinction of Scotus: the *affectio justitiae* – the desire for the good in itself. The reason why the sociobiologists see the self-perfecting tendency of nature as selfish and, therefore, pejorative is that they see this tendency as the *only* tendency in humans. Scotus, on the other hand, sees the good of self-perfection as only *one* possible good.

The *affectio justitiae* is the source of true freedom or liberty of the will, as well as being a restraint on the *affectio commodi*. The *affectio justitiae* is the means by which we can transcend nature and go beyond ourselves and our individually defined good. The *affectio justitiae* is what allows us to go beyond ourseves to see the value of another being.

> To want an act to be perfect so that by means of it one may better love some object for its own sake, is something that stems from the affection for justice, for whence I love something good in itself, thence I will something in itself (*Ordinatio* IV, suppl. dist, 49, qq. 9–10 in [21], p. 185).

Allan B. Wolter notes four characteristics of the *affectio justitiae*. First, it gives us the capacity to love a being for itself rather than for what it can do for us. Second, it enables us to love God for who God is rather than for the consequence of God's love on us. Third, the *affectio justitiae* allows us to love our neighbor as ourselves thereby making each individual of equal value. Finally, such a seeking for the good in itself leads to a desire to have this good beloved by all, rather than being held to oneself ([22], p. 151). This leads Wolter to the conclusion that the *affectio justitiae* amounts to a "freedom *from* nature and a freedom *for* values" ([22], p. 152). Or, as Scotus puts it:

> From the fact that it is able to temper or control the inclination for what is advantageous, it follows that it is obligated to do so in accordance with the rule of justice that it has received from a higher will (*Reportatio Parisiensis* II, d. 6, q. 2, n. 9 in [22], p. 152).

Such an understanding of will as *affectio justitiae* frees the will from the constraints of the necessity of a nature's act of self-realization or the seeking of its own good only. For Scotus then, when a free agent acts according to nature to realize itself or to seek its own good it paradoxically acts *unnaturally* since to seek what is "*bonum in se* is not to seek something that 'realizes the potential of a rational nature.' It is somehow to transcend 'the natural' and thus to have a mode of operation that sets the rational agent apart from all other agencies" [3].

This understanding of will grounds, for Scotus, the possibility of our being able to transcend our own self-interest or what sociobiologists call genetic selfishness. And this is the foundation of human freedom.

Of critical significance is that the concept of freedom that Scotus proposes is not limited to choice of alternatives or freely elicited acts. Rather, in keeping with his mentors Augustine and Anselm, Scotus views freedom as "a positive bias or inclination to love things objectively or as right reason dictates" ([22], p. 152). That is, the proper focus of freedom, and by implication moral analysis, is not the individual act of choice, but the inclination as a whole. And such an inclination focuses on fidelity to the good in itself, not the specific act of choosing that good nor the necessary appreciation of what is good for the fulfillment of the nature of the agent.

Let us examine Scotus' concept of freedom to see more clearly its relation to the *affectio justitiae*. From an experiential perspective, Scotus distinguishes three types of freedom. First is the capacity to choose contrary acts: to continue to sing or not to sing. Second we have the capacity to choose contrary objects, to sing or take a nap, each of which could be done, but not simultaneously. Third is the ability to choose opposite effects, to sing an aria or a ballad. But regardless of what one does, at the core of each of these acts is the experience that

> at the very moment that it wills or causes something, it could equally well will the contrary. A decision of the will never takes away its potentiality to act in the opposite way ([5], pp. 148–149).

The core issue here is that of contingency, the fact that the opposite could just as well have been chosen as that which we actually chose. Thus, whenever we will something, Scotus argues, we also experience that we could have just as well not have willed it. Such choice also extends to the goods we experience, for we can be offered a good and know that this is a good and yet refuse it. And for those still not convinced of the contingency of our acts, Scotus suggests,

> And so too, those who deny that some being is contingent should be exposed to torments until they concede that it is possible for them not to be tormented (*Reportatio Parisiensis* I, pro., q. iii, art. 1 in [23], p. 9).

In his discussion of freedom, Scotus follows the older Catholic tradition coming from Augustine through Anselm, who said "Whoever has what is appropriate and advantageous in such a way that it cannot be lost is freer than he who has this in such a way that it can be lost" ([1], p. 378). From a psychological point of view, Scotus argues that our awareness of the limitation of any particular act of will means that we experience freedom as choice. That is, we are aware that we could have chosen otherwise and that such a choice would have given a different degree of perfection. Thus, "choice is simply basic freedom in inferior conditions" ([9], p. 87), i.e., human finitude. When willing, our will is never fully actual for it is contingent. Yet for all that, we can approach our perfection through our steadfastness or constancy in cleaving to the object of our love. "The perfection of freedom connotes a perseverance and stability in the will's adherence to the good" ([9], p. 98).

Scotus presents both a critical and a positive perspective on freedom that is of particular importance. He discounts the significance of choice, understood as any particular choice or as any choice considered as an isolated event. To say this, of course, flies totally in the face of certainly the normative American experience of freedom and perhaps the Enlightenment tradition as well. For we revel in individual choice and assume that this is the essence of freedom. Such freedom is the core of autonomy, our expression of self-determination. From early in our lives, we are taught that ahead of us lies a series of decisions which will shape our lives and for which we alone will be responsible. As Americans, we have taken to heart existentialism's perspective that our existence precedes our essence and that one becomes one's self only through particular, individual choices. And if such choices are absent, one remains unauthentic.

Scotus fashions his critique from a theological perspective which grapples with the question of how God can be free if love for the divine essence – for only an infinite being can fulfill an infinite being – is necessary. Scotus develops two formulations of freedom to respond to this. The first looks to love for finite objects and is the ability "*not to limit* oneself to limitedly perfecting objects" ([9], p. 83). The second envisions love for God and freedom is the "ability to *continually adhere* to the unlimitedly perfecting object ([9], p. 83). The point common to both formulations is the will's ability to achieve perfection "through active union with its beloved" ([9], p. 83). This holds true regardless of whether the will is infinite and *de facto* there is no other intentional object or whether the will is finite and there are multiple intentional objects. Thus for Scotus the essence of freedom is not choice but what he calls *firmitas* or what we could call fidelity or constancy.

What follows from this is that the finite will can never fully express its basic freedom because for humans there will always be another intentional object, another "what if I would have done this?" which could lead to another version of ourself. Thus for us to choose one goal is to abandon others together with the perfection they could have given us. And given that we are finite, it is not necessary – as it is for God – to choose that which would ultimately perfect us. Freedom thus manifests itself in choice: "basic freedom in inferior conditions" ([9], p. 87), that is, in the context of finitude.

For Scotus, however, free will is not limited only to the fact of choice or even appropriately characterized by it. Rather, choice is "reflective of a deeper structure at work in a specific situation" ([9], p. 85). And this deeper structure is steadfastness which constitutes the perfection of the will: "a perseverance and stability in the will's adherence to the good" ([8], p. 77).

The second critical issue is Scotus's distinction between the *affectio commodi* – the inclination to do what is to my advantage – and the *affectio justitiae* – the inclination to do justice to the intrinsic reality of a particular being or situation. In the former, we have nature seeking its self-perfection, whether in the individual or the species. Seeking what is to one's advantage is seeking to actualize the potential instilled in it by virtue of what the being is. Seen as a nature, this quest for perfection is a good because it fulfills the nature by enabling it to become what it is. Thus, Scotistic thought would agree with the sociobiologists that as natures we, like any other nature, seek our good and our individual perfection and that we do so necessarily. But it would disagree that this is selfish in the pejorative sense of sociobiology. In fact, I think from a Scotistic perspective, the sociobiologists discussion of genetic selfishness makes no sense at all and is a significant distortion of human existence.

The affection for justice is the capacity to love something or someone for their own selves, regardless of whether this happens to be a good for me or not. As Wolter phrases it, this is a "freedom *from* nature and a freedom *for* values" ([22], p. 152). The conclusion is the paradox that

> what differentiates the will's perfection as nature from the perfection of all other natural agents is that it can never be attained if it be sought primarily or exclusively: only by using its freedom to transcend the demands of its nature, as it were, can the will satisfy completely its natural inclination ([22], p. 154).

Scotus's affirmation here is that we have the capacity to value an entity for its own sake, independent of its personal or social utility. As Scotus would phrase it, we have the ability to transcend the capacity to do justice to ourselves by doing justice to the good itself. The strong claim is that we are capable of recognizing goods distinct from our self-perfection and independent of our interests and choosing them even though such a choice may run counter to our personal self-interest or what does justice to my own nature.

> The will by freely moderating these natural and necessary tendencies to happiness and self-perfection is able to transcend its nature and choose Being and Goodness for their own sake. . . . Thus the free will is not confined to objects or goods that perfect self, but is capable of an act of love. . . . love is the most free of all acts and the one that most perfectly expresses the will's freedom to determine itself as it pleases ([13], pp. 630–631).

The conclusion is that one can distinguish at least a good and a better in human life. What is good in human life is a life that perfects us, that brings our being to a greater actualization. This is the realization of the *affectio commodi*. But what is better is the transcendence of self either to appreciate goods independent of us or even to curb our legitimate interest in self-perfection to seek the good of others for their own sakes. This is the realization of the *affectio justitiae*.

Put existentially,

> A free choice, then is the meaning of existence and the total initiative is left to man to rightly moderate his natural tendencies in the pursuit of being for its own sake. And in this sense one's existence is one's own responsibility and depends on one's causal initiative as an ultimate response to Being or Nothingness ([13], p. 631).

Put ethically,

> right reason also recognizes that our self-perfection, even through union with God in love, is not of supreme value. It enables man, in short, to recognize that the drive for self-perfection paradoxically must not go unbridled if it is to achieve its goal, but must be channeled lest it destroy the harmony of the universe intended by God ([22], p. 153).

What is most helpful about this perspective is that while it affirms self-perfection, ultimately such a perfection is not an end in itself. To "be all that we can be" we must step beyond the confines of self and actualize that most free of all acts, an act of love. For only then do we find ourself open to the depths of reality. And in the steadfast adherence to that beloved, we realize the fullness of freedom.

V. Conclusion

The burden of this paper has been to argue for the rejection of three core claims of sociobiology: scientific materialism, genetic selfishness, and determinism. In this conclusion I focus my claims and conclude with a metaphor more appropriate for expressing human agency in the context of modern genetics.

To do this I will discuss three experiences that we humans have as ways of re-focusing the arguments made above.

A. *Human Concern for Other Species*

I will discuss this issue from the perspective of a recent article discussing sociobiology in relation to Catholic ethics. Stephen Pope describes a sociobiological definition of altruism as an "action which contributes to the fitness of another person at the expense of one's own fitness" ([14], p. 268). Two types of altruism are presented: kin which is the giving of preference to members of one's genetic lineage and reciprocal which is a special preference for those with whom one has a reciprocal relation ([14], p. 268). Opposed to this is "promiscuous altruism, i.e., altruism practiced 'without discrimination of kinship, acquaintanceship, shared values, or propinquity in time or space'" ([14], p. 274). Such altruism is not possible according to the laws of evolutionary process.

While Pope argues against both a biological reductionism and a disembodied personalism in trying to ground an ethic of love and altruism within a sociobiological understanding of human nature, I think the sociobiological project runs into some difficulty. For example, Pope notes that "interpersonal reciprocity [investments in a social network from which one will benefit over the long haul] involves attitudes and affections that cannot be fully understood or completely generated by biological features of human nature, without remainder" ([14], p. 281). Yet the very next sentence observes: "The capacity for mutuality must of course lie within and to some extent depend upon our biological natures, otherwise we would be 'disembodied spirits'" ([14], p. 281). How such a value arises is not explained by sociobiological theory. The difficulty of this account is a superimposition of a value set onto what seems to be, from a sociobiological perspective, biological givens. However the tension in the two sentences quoted above is, from my perspective, not fully resolved. Additionally, while the project of engaging in a dialogue with the ethical tenets of sociobiology is worthwhile, it is unclear why one must accept its definition of altruism, especially when there are so many *prima facie* difficulties in the concept as defined by sociobiology as noted above and between it and the Christian tradition.

The previously discussed Scotistic distinction between nature and will pushes this discussion in an interesting direction. Nature, for Duns Scotus, is the disposition or appetite to seek self-perfection. That is, a nature is the inclination of a being to seek its own good or the good of its kind. It is the principle of activity by which a being actualizes its reality, or the reason a being acts as it does. Nature, for Scotus, is the answer to the question: "Why does this being act as it does"?

Thus, to put this into sociobiological terms, when an animal – a dog or a lion, for example – acts to protect its offspring, it is seeking to achieve the good of its species, i.e., preservation. It is attempting to guarantee a place for its descendants in the world. Through actualizing its nature in this way it is realizing a good and, indeed, perfecting itself. What sociobiology calls kin altruism Scotus calls acting according to nature or seeking to do justice to one's self. Seeking the good of those with whom one has a special relation – through marriage or friendship, for example – also brings a certain perfection to one's self. Scotus argues that a nature as such has no choice but to seek its good. That is, a nature seeks its perfection of necessity. That is what it is programed to do, in our contemporary idiom.

In Scotus' perspective, what differentiates the human being from other beings is that the human also possesses the capacity for self-transcendence, what he calls the will or the capacity to do justice to some other being or value. The will, understood in this sense, is the capacity to value something for its own sake. Scotus argues that as humans we have a two-fold capacity: a capacity to seek our own perfection or good and a capacity to transcend the boundaries of self and appreciate the good of another for its own sake.

This argument of Scotus explains what sociobiology cannot: human actions to protect another species, even at a cost to the human. Arguments, for example, about animal rights, the protection of endangered species, or the ecological movement itself cannot fully be explained on the basis of self-interest, enlightened though it may be. Such movements require that the natural tendency to self-perfection in humans be checked in some way so that beings of another species or the earth itself can be protected. Altruism understood in a sociobiological perspective cannot ground easily such activities on the part of one species for another.

For Scotus, the capacity to give another its due or to recognize the value of another entity apart from its value to me is strong evidence for an ability to transcend one's self – what he calls the native freedom of the will. Thus, the Scotistic perspective on the will – as distinct from nature – helps us appreciate the full range of human capacities as well as going a long way to help explain why and how it is that humans can value the good of others as much or even more than their own.

B. *Human Love*

Another experience to examine is human love. Sociobiology puts severe constraints on this experience by defining it primarily in terms of reciprocity among those with whom I share genetic identity or in terms of my genes attempt to maximize their presence in the gene pool. Other efforts at community are described exclusively in biological/Hobbesian terms of self-interest and self-preservation. As Dawkins says,

> Be warned that if you wish, as I do, to build a society in which individuals cooperate generously and unselfishly towards a common good, you can expect little help from biological nature. Let us try to *teach* generosity and altruism, because we are born selfish. Let us understand what our own selfish genes are up to, because we may then at least have the chance

to upset their designs, something which no other species has ever aspired to ([6], p. 3).

Two questions arise from this citation. First, whence comes the desire for such behavior? Second, if such behavior is so contrary to our nature, on what basis will such teaching be grounded, much less be successful? The Scotistic perspective developed above gives us a way into these questions.

The desire for such generosity and altruism is a part of our experience. And it arises, negatively stated, from our ultimate dissatisfaction with either functionalism or utilitarianism as our primary or exclusive mode of relating to either objects or persons. Now clearly we all relate to objects and persons functionally. We do put these entities in means-to-end relations. Baldly stated we use people and objects for our own purposes. We also know that if such functional relations are the primary mode of relating with other persons that such relations soon cease. Others simply refuse to let themselves be treated in such a way. And we too know that such a mode of relation is not ultimately satisfying.

Even with respect to the objects in our lives, we often look beyond pure function to the ascetic dimension of the object. While one flashlight might be interchangeable with any other and serve the same basic function, we tend to prefer those better designed to fit the hand. While one table might just as well support our plates, one with a certain elegance of design tends to be preferable.

As Duns Scotus would phrase it, and this is the positive expression of the experience, we have the capacity to value a good in itself or for itself. That is, we do experience both objects and people as having a value or meaning independent of my own needs, desires, or self-interest. We do experience others as worthwhile in themselves.

A central experience of this is an act of love through which we touch and experience the other person as valuable in their own right. We experience value independent of our own needs. And we experience a desire to share such a good with others precisely because we see the good of this person in his or her own right independently of my own interests. If there were no value of persons apart from their utility to me, we would not want others to meet our friends or our lovers. We do this precisely because we see a beauty or good that transcends our interests.

Scotus calls this the *affectio justitiae*, the capacity to recognize a value or a good for its own sake. It is also the basis for the capacity to control our personal behavior.

> Therefore, this affection for justice, which is the first checkrein on the affection for the beneficial, inasmuch as we need not actually seek that towards which the latter affection inclines us, nor must we seek it above all else (namely, to the extent to which we are inclined by this affection for the advantageous) – this affectio for what is just, I say, is the liberty innate to the will, since it represents the first checkrein on this affection for the advantageous ([21], pp. 469–470).

Again one needs to reflect on one's experience. We know that we are not passive subjects of every desire or emotion that comes our way. Minimally we know that we have the capacity to delay the satisfaction of some desires. We know that we put our desires and wants in some sort of priority. But we also know that we can put a good of some one else above our own. This is the nature of parenting, for example. We know that we can refuse some personal satisfaction in the name of some other good. This is a foundational experience for appropriate interpersonal relationships. And we know that we can value objects above our own interests. This is one of the bases, for example, of the ecology movement. Our experience shows us the evidence of what Scotus calls the *affection justitiae*. And its living presence in us is a refutation of a core claim of sociobiology – namely, that we cannot act beyond our own good which good must itself be defined only genetically.[2]

Important to note here is that, just as sociobiology argues, this *affectio justitiae* is not a motive. It is a behavior. We do in fact act this way. The basis for such action can be found in the previous discussions of the capacity for surprise within evolution by Eaves and Graves and Rahner's discussion of matter. Both of these provide the experiential and empirical foundation for Scotus' elaboration of our capacity to do justice to a good or value other than our own selves.

C. *Human Agency: Transcending the Genome*

Given Wilson's scientific materialism and his sense of the limits of cultural development just presented, what can we expect for the future? What changes can we anticipate?

1. *Wilson and Cultural Development*

Wilson says that there is "a limit, perhaps closer to the practices of contemporary societies than we have had the wit to grasp, beyond which biological evolution will begin to pull cultural evolution back to itself" (18], p. 79). Even, he says, when self determination is turned on full-blown and all options are available, only a few directions will be desirable. "Others may be tried, but they will lead to social and economic perturbations, a decline in the quality of life, resistance, and retreat" ([18], p. 79). Thus for Wilson the question of interest is the extent to which the "hereditary qualities of hunter-gatherer existence have influenced the course of subsequent history" ([18], p. 88). Wilson's answer is that the influence has been substantial.

Thus the emergence of civilization has followed a definable sequence and has led to parallelism in the major features of civilizations. The key to this is hypertrophy: "the extreme growth of pre-existing structures" ([18], p. 89). This means that:

> the basic social responses of the hunter-gatherers have metamorphosed from
> relatively modest environmental adaptations into unexpectedly elaborate,
> even monstrous forms in more advanced societies. Yet the directions this

change can take and its final products are constrained by the genetically influenced behavioral predispositions that constituted the earlier, simpler adaptations of preliterate human beings ([18], p. 89).

Although Wilson argues that knowledge, primarily from science and technology, can equalize people and break some barriers, it cannot "change the ground rules of human behavior or alter the main course of history's predictable trajectory" ([18], p. 96). Or as Wilson phrases it,

Now there is reason to entertain the view that the culture of each society travels along one or the other of a set of evolutionary trajectories whose full array is constrained by the genetic rules of human nature ([18], p. 207).

The only openness to radical change comes from our descendants who may learn how to change the genes themselves. Genetic engineering can be the basis of change with cloning the means to rapid selection. Given that genetic engineering gives us the capacity to change our nature, the question Wilson is left with – and for which he provides no answer – is what we will choose.

2. *The Gene and Human Agency*

If it is the case that humans are the repository of genes seeking to replicate themselves and that our future is shaped by our past, it would seem that we are at the mercy of our genes, that talk of human agency – even through genetic engineering – would simply be that – talk. The previous two sections of this conclusion already made counter-arguments against this position of sociobiology. In this final section I speak to the issue of human agency directly, first from within the context of genetics and then from a discussion of intentionality.

At the beginning it is important to note that this discussion will occur within the context of contemporary genetics. That is, human agency will not be presented acontextually. The image of a pure self, independent from either genes or culture, is one that I take as false. Additionally such an image has a bad press in general and has been rightly subject to appropriate philosophical critiques. Finally this image of the radically autonomous self seems to be the one lurking behind discussions of agency in sociobiology and mechanistic thought in general. Thus, the concept of agency needs a measure of clarification greater than what can be provided here. My agenda is to examine the concept of agency within the framework of contemporary genetics. R. David Cole quite crisply phrases the approach I wish to take in speaking of the ways in which human agents might "transcend genetic determinism to develop their lives within a broad range of predetermined potentials" ([4], p. 2).

Three processes specify how information from the genome sets the phenotype of the organism. First is a three-fold process of replication of the sequence of nucleotide bases from the DNA, the transcription into a nucleotide sequence by the RNA, and the translation of the base sequences into proteins.

([4], p. 3). Second, is a pattern of specification of "patterns of information inherent in the sequence of bases in DNA" ([4], p. 3). Finally, there is "the capability of the patterns of interaction to be modulated by forces external to the cell and beyond direct control by the gene" ([4], p. 3). This third level is significant for it is here that the actual expression "of (potential) genetic information becomes contingent" ([4], p. 4). Thus, even though the range of contingencies are limited, there are possibilities of variation and even independence in expression. As Cole states it: "The genome determines all the potential states of being and behavior for the organism, but it does not predetermine the organism to any one particular state" ([4], p. 6).

Thus, how our genes express themselves is contingent on the process of development through which all organisms pass as well as the environment in which this development occurs. Specifically, each "successive adaptation is superimposed on its predecessor so that a terminally differentiated cell manifests the entire history of its cell line and not merely its immediate state" ([4], p. 8). This differentiation is essentially irreversible and means that "the creature designed by the DNA has come to control the expression of its designer" ([4], p. 8). The consequence of this process is that "as the author of our biological destiny, DNA has been reduced to the role of co-author, even if it might be the senior author" ([4], p. 8).

Cole argues that the genome sets a range of potentials that are both internal and external to the genome. Within those boundaries there is an area of contingency in which "we have some control over our destiny" ([4], p. 13) and thus "free will could synergise with our genetic system to maximize some potentialities and minimize others" ([4], p. 13). Thus, within the very biological structure of the genome there is an empirical foundation for the experience we have of our own agency. Critical in Cole's presentation is the metaphor of co-authorship. And what is critical in this metaphor is the relation of both constraint and initiative. On the one hand, the genome sets limits on the possibilities of expression. But on the other hand, "genetically based systems of cells develop and differentiate by a series of adaptations into a creature that exerts control over its creator, DNA . . ." ([4], p. 13).

3. *Change as Co-Creation*

Another metaphor to evoke the concept of human agency is that of created co-creator as presented by Philip Hefner [10]. Similar to the metaphor of co-author, the created half of the metaphor focuses on our dependence. This dependence is two-fold. First, it is biological in that we evolve and are dependent on our past. Second, from a religious perspective we are radically dependent in that our very being is dependent on the "creative grace of God" ([10], p. 225). Thus, though we need not exist at all, given our existence, we exist only within the context of our natures which have arisen from the evolutionary process.

The creator half of the metaphor, corresponding to the author image, recognizes our ability to use "our cultural freedom and power to alter the course of historical events and perhaps even evolutionary events" ([10], pp. 225–226). Yet here too there is a dependence: the evolutionary past out of which we yet emerge. But there is also an openness: "We make decisions and take actions that determine in part the course of events. Events bring new things. The human race is daily inventing new things that hitherto never existed" ([10], pp. 227–228). Robert John Russell articulates this image in an evolutionary context by suggesting that:

> the action of humans in biological adaptation will be an instance of intervention by a higher level cause, an instance of a "miracle" to the DNA/RNA/proteins at hand, and thus a model for us of how God might act in the dimensions of spirit and history by what seems to be intervention to us, but is creaturely instrumentality to God ([16], p. 18).

4. *Intentionality and Agency*

The experience of intentionality also presents an important perspective on human agency. John Searle locates the issue this way:

> . . . it is in part because the Darwinian revolution was successful in showing that the appearance of mental intervention in the development of species could be completely accounted for on mechanical hypotheses that some sociobiologists were erroneously led to the conclusion that mental states play no causal role in the explanation of the behavior of specific animals within the various species. ([17], p. 167).

Searle's project is to force a separation between "the hypothesis of non-teleological, nonintentionalistic mechanisms of natural selection from the hypothesis of nonteleological, nonintentionalistic explanations of animal behavior" ([17], p. 168). His argument is that while the former is correct, the latter is mistaken for "the explanation of large areas of human behavior and presumably large areas of other animal behavior is ineliminably intentionalistic in form" ([17], p. 168).

One part of Searle's argument is that while the specification of a desire gives a causal account of what the actor does, "in some cases it [the causal account] does not determine the action. The agent determines the action. He decides to act on the desire or not" ([17], p. 174). In removing acting on desire from a Newtonian mechanistic explanation and describing such acts as intentional, Searle echoes Scotus's analysis of free acts as uncaused causes. As Scotus rejected the Aristotelian axiom "Quidquid movetur, ab alio movetur" as a contradiction of freedom – for if the act of will were caused it would no longer be free – so too Searle rejects Newtonian mechanics as an explanation of some human actions. Searle differs from Scotus in that he assumes that some reason causes us to act, but he agrees that a mechanistic account of human acts fails to do justice to the reality of the situation.

Searle further argues that the act of communication mandates a stronger sense of intentionality. Wilson's definition of communication cited by Searle is: "Biological communication is the action on the part of one organism (or cell) that alters the probability pattern of behavior in another organism (or cell) in a fashion adaptive to either one or both of the participants" ([17], p. 176). Searle thinks it is necessary to distinguish communication "from any other form of biologically mediated transfer of information from one organism to another" ([17], p. 176) for otherwise any information transfer counts as communication.

For Searle communication involves a double intentionality:

> First, there must be an intentional state in the speaker or sender of the signal. Second, the speaker or sender must intentionally perform some act designed to induce in the receiver or hearer an awareness of the speaker's intentional state ([17], p. 177).

Communication is not simply acquisition of information. What distinguishes the two is intentionally: "the intentional transfer of an intentional state from the speaker to the hearer" ([17], p. 178). That is, the receiver does not just pick up information. Rather, the receiver obtains information which reflects a state in the sender and which the sender intends for the receiver to have. The act of communication is purposeful and goal-directed and is distinguished from information acquisition by precisely the transfer of the intentional state from one entity to another. Neither Wilson's definition nor a behaviorist or a mechanistic account do justice to this reality. Agency beyond mere epiphenominism is required to do justice to the reality of communication.

From a genetic, a philosophical, and theological perspective, there is a meaningful way in which human agency can be grounded and affirmed. While the concept of agency described herein is a modest one in that it recognizes the role of the genome within an evolutionary context, it is also a powerful one in that it affirms the transcendence of the human agent from the genone. The images, metaphors, and arguments sociobiology uses are both flawed and inadequate in their attempt to capture the richness of human experience. The scientific materialism proposed by Wilson and genetic selfishness so vigorously promoted by Dawkins have fatal flaws which have been revealed through counter arguments and counter perspectives. These perspectives, though not fully developed here, point to new directions for the development of more adequate accounts of human behavior.

Perhaps the best integrating image for this perspective can be found where we began: in the poetry of Gerard Manly Hopkins ([12], p. 66).

> In a flash, at a trumpet crash,
> I am all at once what Christ is, since he was what I am, and
> This Jack, joke, poor potsherd, patch, matchwood, immortal diamond,
>   Is immortal diamond.

Worcester Polytechnic Institute
Worcester, MA, U.S.A.

NOTES

1. This paper was written with the support of "Theological Questions Raised by the Human Genome Initiative". It was sponsored by The Center For Theology and The Natural Sciences at the Graduate Theological Union, Berkeley, CA. National Institutes of Health Grant No. GNM 1 R01 HG00487-01.
2. Such a tendency can also be related to the so-called "preferential option for the poor" in which the goods of other neither genetically nor socially related to an individual can both be chosen and acted upon. This idea cannot be developed here because of lack of time.

BIBLIOGRAPHY

1. Alluntis, OFM, Felix and Wolter, OFM, Allan B. (trans.): 1975, *John Duns Scotus, God and Creatures: The Quodlibetal Questions*, Princeton University Press.
2. Boehner, OFM, Philotheus: 1955, 'The Teaching of the Sciences in Catholic Colleges', Franciscan Educational Conference, pp. 150–159.
3. Bowler, John: 'The Moral Psychology of Duns Scotus: Some Preliminary Questions', *Franciscan Studies*, Forthcoming.
4. Cole, David, R.: 1992, 'The Molecular Biology of Transcending the Gene', *Pro manuscripto*.
5. Creswell, J.R.: 1953, 'Duns Scotus on the Will', *Franciscan Studies* 13, 147–158.
6. Dawkins, Richard: 1976, *The Selfish Gene*, Oxford University Press, New York.
7. Eaves, Lindon and Gross, Lora: 1992, 'Exploring the Concept of Spirit as a Model for the God-World Relationship in the Age of Genetics', *Zygon* 27, 261–285.
8. Frank, William A.: 1982, 'Duns Scotus' Quodlbetal Teaching on the Will', PhD. Dissertation, Catholic University of America.
9. Frank, William A.: 1982, 'Duns Scotus' Concept of Willing Freely: What Divine Freedom Beyond Choice Teaches Us', *Franciscan Studies* 42, 68–89.
10. Hefner, Philip: 1989, 'The Evolution of the Created Co-Creator', in Ted Peters (ed.), *Cosmos as Creation: Theology and Science in Consonance*, Abingdon Press.
11. Hopkins, Gerard M.: 1963, 'God's Grandeur', in W.H. Gardner (ed.), *Poems and Prose of Gerard Manley Hopkins*, Baltimore: Penguin Books.
12. Hopkins, Gerard M.: 1963, 'That Nature is a Heraclitean Fire and of the Comfort of the Resurrection', in W.H. Gardner (ed.), *Poems and Prose of Gerard Manley Hopkins*, Baltimore: Penguin Books.
13. Messerich, OFM, Valerius: 1968, 'The Awareness of Causal Initiative and Existential Responsibility in the Thought of Duns Scotus', *De Doctirna Ionnis Duns Scoti*, Vol. II, Problema Philosophica, Acta Congressus Scotistici Internationalis, Rome, pp. 629–644.
14. Pope, Stephen: 1991, 'The Order of Love and Recent Catholic Ethics: A Constructive Proposal', *Theological Studies* 52, 255–288.
15. Rahner, S.J., Karl: 1983, 'Christology Within An Evolutionary View', Kruger, Kark-H. (trans.), *Theological Investigations*, Crossroad, New York, vol. 5, pp. 157–192.
16. Russell, Robert John: 1992, 'Seekers and Makers of the Divine Will: Theology in the Context of Molecular Biology', *Pro Manuscripto*.
17. Searle, John R.: 1978, 'Sociobiology and the Explanation of Behavior', in Michael S. Gregory, Anita Silvers and Diane Sutch (eds.), *Sociobiology and Human Nature*, Jossey-Bass Publishers, San Francisco.
18. Wilson, E.O.: 1978, *On Human Nature*, Harvard University Press.
19. Wilson, E.O.: 1980, *Sociobiology: The Abridged Edition*, The Belknap Press of Harvard University.
20. Wolter, OFM, Allan B.: 1961, *Select Problems in the Philosophy of Nature*, *Pro Manuscripto*, St. Bonaventure University, The Franciscan Institute.

21. Wolter, OFM, Allan B. (eds.): 1986, *Duns Scotus on the Will and Morality*, The Catholic University of America Press.
22. Wolter, OFM, Allan B.: 1990, 'Native Freedom of the Will as a Key to the Ethics of Scotus', in Marilyn McCord Adams (ed.): 1990, *The Philosophical Theology of John Duns Scotus*, Cornell University Press.
23. Wolter, OFM, Allan B. (trans.): 1987, *John Duns Scotus: Philosophical Writings*, Hackett Publishing Co., Indianapolis.

ROBERT LYMAN POTTER

A Comparative Appraisal of Theocentric and Humanistic Ethics Systems in the Clinical Encounter

INTRODUCTION

I claim that a theocentric perspective is more acceptable than a secular humanistic perspective as a support for the ethical principle of respect for the autonomy of persons. My argument is in four parts: (1) the principle of respect for the autonomy of persons entails both a description of persons and a meta-ethical foundation for the concept of persons; (2) either a secular humanistic or a theocentric conceptual system can be used to describe and ground the concept of persons; (3) the comparative appraisal method of Imre Lakatos has the explanatory power to organize a test as to which of the conceptual systems is the more acceptable, and, therefore, the more progressive; (4) after doing a comparative appraisal I have concluded that James M. Gustafson's theocentric system is more progresssive than H. Tristam Engelhardt's secular humanistic system as a support for the ethical principle of respect for the autonomy of persons.

The warrant for Gustafson's system being more acceptable is the inclusion of the concept of human fault in the theocentric account of persons. The inclusion of human fault in the constitution of persons results in a more realistic expectation of the extent to which the principle of respect for autonomy can be fulfilled in the clinical encounter.

ENGELHARDT'S CONCEPT OF PERSON

In Engelhardt's system of ethics the concept of person is central. "The principle of autonomy and its elaboration in the morality of mutual respect applies only to autonomous beings. It concerns only persons. The morality of autonomy is the morality of persons" ([5], p. 108).

Engelhardt defines "person" according to three specifications: self-consciousness, rationality, and a sense of moral concern. The first criterion, self-consciousness, entails the fundamental notion that the human species has evolved to the level at which self-reflection is a capacity of each fully

203

E. E. Shelp (ed.), Secular Bioethics in Theological Perspective, 203–217.
© 1996 Kluwer Academic Publishers. Printed in the Netherlands.

developed member of the species. This self-reflective capacity, unique to humans, may be seen in preliminary degrees in some other animal species. However, only in humans is the self-reflective capacity fully potential, and only in persons is it fully realized.

Only members of the human species who have developed and sustained a sufficient central nervous system will exhibit this self-reflective capacity. According to Engelhardt, "fetuses, infants, the profoundly mentally retarded, and the hopelessly comatose" do not have this minimal central nervous system capacity to support the self-reflective function ([5], p. 107). Without an adequate degree of self-consciousness, a human being is not qualified to be a person.

According to the second criterion, in order to qualify as a person a human must demonstrate the capacity for rationality. The demarcation line between rationality and irrationality is not clarified in Engelhardt's model. Rationality may be taken to mean that a person at some time, but not necessarily all times, demonstrates an incomplete, but acceptable, capacity for deliberation according to logical information processing. The range of humans qualifying for personhood will depend on the definition of the standards of "acceptable". If the norm of acceptability is very broad then a larger portion of humans will be qualified as persons. How far outside an acceptable norm society will allow an individual to fall and still be counted as a person is the unresolved question which is adjudicated in court rooms, estimated at the bedside, and puzzled over in the everyday world.

The third criterion, possession of a moral sense, is also difficult to standardize as a qualifying norm. Moral sense, in Engelhardt's model, is the product of "an inner life with some prereflective anticipations of values as values are understood by self-conscious free agents" ([5], p. 112). Persons can compare one value with other competing values and choose among values. This comparison is a process in which appreciation is a key concept. Appreciation allows for distinctions among values. These distinctions ground "the notion of worthiness of blame and praise: a minimal moral sense" ([5], p. 107).

To summarize the core of Engelhardt's anthropological model, persons are those humans who demonstrate acceptable degrees of capacity for self-consciousness, rationality, and moral sense.

PERSONS NETWORK INTO MORAL COMMUNITIES

Persons tend to network themselves into moral communities where a value system held in common becomes the symbolic organizer of a moral consensus. For Engelhardt, "the very notion of a moral community presumes a community of entities that are self-conscious, rational, free to choose, and in possession of a sense of moral concern" ([5], p. 105). It is the moral community which provides the content-full moral vision by which the member persons construct their own guides to conduct.

The tendency for persons to network themselves into moral communities with traditions is in part the origin of what Engelhardt calls "the bifurcation of the moral life" ([5], p. 105). The bifurcation consists of the tendency of persons to agree with moral friends who share a moral vision, and, at the same time, the need to live and work with moral strangers who do not share a moral vision. "On the one hand, individuals live their lives within particular concrete, content-full moral visions. On the other hand, insofar as they live in a tolerant, secular pluralist society, they will recognize that many do not understand the true character of concrete morality. In the impoverished moral sphere where moral strangers meet, much must be tolerated that one deeply abhors and condemns" ([5], p. 124). This bifurcation is a constant and necessary feature of life in a pluralistic society. To successfully live with both friends and strangers a special neutral moral space must be created where differences can be peacefully negotiated.

NEGOTIATION AMONG MORAL STRANGERS

Since the Enlightenment, there has been a strong expectation that a liberal, rational, and democratic state would result in the establishment of a content-full moral vision. The content-full moral vision was to be the cultural product of enlightened secular humanistic reason. This consensus was to be the one sufficient guide to good conduct and, thus, eliminate moral fragmentation and all other conditions which create moral strangers. Engelhardt concludes that secular humanism cannot produce and support a canonical, moral account full of content ([5], p. 110).

Having concluded that secular humanism has failed to produce a content-full moral vision which can persuade all members of the larger community, Engelhardt realizes that some adequate substitute for this lost consensus must be found. Since the project of cultural consensus had failed, except among moral friends within a particular moral community, it was necessary to develop a substitute project of negotiating differences into some accommodation which moral strangers can accept.

He concentrates on what persons must do as moral strangers in a morally fragmented world to attempt to create a morally neutral space in which to bring divergent moral traditions into a peaceable kingdom where negotiation of differences can be undertaken. Especially important for Engelhardt is that persons must have the moral will and the capability to create such a neutral moral space. Only persons who are capable of exercising their self-consciousness, rationality and moral sense can agree to enter into the negotiation toward creating a peaceable kingdom. Within this neutral moral space, or peaceable kingdom, the conflicted positions may be negotiated by the persons involved in the relationship to the mutual satisfaction of each person in the relationship.

MORAL STRANGERS IN THE CLINICAL ENCOUNTER

The relationship of the patient and physician might be characterized either as the relationship of moral friends or of moral strangers. Engelhardt is concerned mainly with the problems of the patient-physician relationship as a clinical encounter of moral strangers. In this case any resolution of moral controversy between the patient and physician can be expected to follow along four lines of conduct: "(1) they can be resolved by appeal to force, (2) an appeal can be made to a set of commonly held moral beliefs, (3) an appeal can also be made to rational arguments, or (4) there is the possibility of negotiating a peaceable resolution to a controversy" ([4], p. 256).

In order to negotiate a peaceable resolution to a clinical controversy there must be a mutual respect demonstrated as a strong adherance to the doctrine of informed consent. "Again, this is why free and informed consent becomes the paradigm ethical issue in a secular pluralist society ... The concrete fabric of the moral life and of the physician-patient relationship in a secular pluralist society is therefore to be invented by appeal to a procedure of mutual respect of the freedom of the participants in the dispute" ([4], p. 257).

From the perspective of the patient, Engelhardt recognizes three senses of freedom for choosing to consent: "(1) being able to choose, (2) being unrestrained by prior commitments or justified authority, and (3) being free from coercion" ([5], p. 266). Engelhardt agrees that there are some individuals who will not qualify for "being able to choose". Restrained freedom is characterized by those who have signed contracts or who have been imprisoned. The use of coercion through threats of violent force or from deception in the information given is unjustified, but not beyond likelihood.

The standard of respect for persons requires that freedom to choose and full disclosure of information will be met in the situation where moral strangers possessing the will to morality negotiate a peaceable settlement. "As such, the core of ethics becomes that of mutual respect of the moral agents involved in a controversy" ([5], p. 257).

THE HUMAN AS THE ULTIMATE GROUND OF MEANING

Engelhardt does not specifically reject an ultimate metaphysical foundation for his concept of persons. Rather than to make any reference to the transcendent, he chooses to ground his ethical system on the basis of human nature. "All of this is to be done without grounding in any ultimate value system other than the humanum itself. This is a reorientation in terms of what one might call a transcendental mooring: persons as the center and source of meaning" ([5], p. 376).

By making this choice Engelhardt clearly prefers to construct an anthropocentric rather than a theocentric ethical system. "Secular humanism in this sense is a humanism that attempts to understand the normatively human and to ground a vision of moral and political theory without reference to, but without rejection of, the supernatural, transcendent, or metaphysical" ([6], p. 90).

In several areas of his writing Engelhardt has demonstrated an openness to the idea of a transcendent or metaphysical ultimate. He has a developed notion of God which is not brought into the core of his ethics of a peaceable kingdom. Instead of God, the humanum constitutes the core of Engelhardt's ethical system.

GUSTAFSON'S CONCEPT OF PERSON

Gustafson shares a considerable agreement with Engelhardt's model of the human. Gustafson uses a critical method to interrogate the human sciences for the most productive and coherent elements from which to model the human being [11]. Biology and the evolutionary conclusions of sociobiology are blended with interactional psychology to create an overview of the human being which is compatible with the strongest currents of scientific thought. Gustafson reaches conclusions similar to those of Engelhardt in regard to issues such as embodiment, self-consciousness, and the desirability of rationality and a minimal moral sense in persons. Gustafson's reliance on interactional psychology corresponds to Engelhardt's use of the idea of a person being embedded in a moral community. Therefore, in large part, the two models being compared are similar in construction.

There are some important differences which distinquish the two models. The main differences are (1) the degree to which a person is a trustworthy autonomous agent, and (2) the theocentric grounding of the respect for persons.

HUMAN FAULT AND THE LIMITS OF AUTONOMY

Central to Gustafson's thought is the concept of human fault: "the capacity for fault is part of our human nature" ([10], p. 294). Gustafson develops an argument which indirectly challenges Engelhardt's confidence that persons can be trusted to be rational and moral in sufficient degree to sustain the peaceable kingdom required to negotiate ethical differences.

Gustafson outlines four types of fault: the experience of misplaced trust or confidence (the traditional problem of "idolatry"), the experience of misplaced valuations of objects of desire (the traditional problem of wrongly ordered love), the experience of erroneous perceptions of the relations of things to each other and of our understanding of things (the traditional problem of "corrupt" rationality), and the experience of unfulfilled obligations and duties (the traditional problem of disobedience)" ([10], p. 294). All four are "inevitable" because of the flawed character of our human nature.

Each fault involves a disordering of the relationships we have to ourselves, to others, to the world around us, or to the "powers that sustain life and bear down upon it" ([10], p. 294). The fault of misplaced trust is disordering of human life by wrongly centering it on a false value. The fault of wrongly ordered objects of desire is the disordering of objects of human valuation and of the wrong intensity of valuation of proper objects. The fault of corrupt rationality is a misconstruing of the reality that engages us. This misconstruing is a disorder of relationship motivated by the kind of interests we hold. The fault of disobedience is a moral flaw which points to the basic disordering "grounded in the characters of agents, not merely in the violation of rules of conduct" ([10], p. 303). The disorder of relationships is the consequence of human fault or "sin".

HUMAN FAULT IN THE CLINICAL ENCOUNTER

Gustafson does not have a written account of his view of the patient-physician relationship. It has to be imaginatively constructed from his model of the faulted human. Gustafson would locate the concept of human fault near the center of the dynamics of the clinical encounter: "the recognition of human finitude, and the self-critical attitude it grounds and informs . . . are indispensable to responsible medical practice" ([9], p. 70). To the extent that human fault pervades all relationships, the patient-physician relationship is distorted by sin.

The four faults Gustafson describes can serve as an outline for the ways in which the patient-physician relationship is likely to be distorted: misplaced trust, misordered desire, mistaken rationality, and misdone obligations.

The first fault of misplaced trust is apparent in the case of a patient not being willing to question the actions of a physician. This also can extend into not being willing to actively participate in decision making in deference to the physician as the primary decision maker. From the perspective of the physician, misplaced trust or self confidence can result in attempting to do more than one's understanding and experience allow.

The second fault of misordered desire is expressed in those self-promoting motives which both physicians and patients commonly hold as their goals. There is no guarantee that either party comes to the clinical encounter free of self-promoting motives.

The third fault of mistaken rationality is the belief that one knows more than they actually do know. Both patients and physicians make this error. The inflated knowledge role of physicians make them particularly vulnerable to actualize this potential fault.

The fourth fault of misdone obligations is witnessed in unfulfilled actions or in damaging actions. Both patients and physicians create conditions for damage when they are not honest and disciplined in their duty.

The concrete outcomes of these distortions will not be detailed here. It is evident that these distortions of human fault are pervasive and persistent in the clinical encounter. If these distortions cannot be corrected or ameliorated then the possibility of peaceful negotiation of the type envisioned by Engelhardt is seriously limited or eliminated all together from the clinical encounter.

Howard Brody's analysis of power in the clinical encounter as well as Jay Katz's description of the silent world of doctor and patient demands that all of the medical scene be looked at in a way which can sort out the faults which have accumulated [2], [13].

Disregard for the autonomy of patients as persons has characterized much of the patient-physician relationship in the past. There is evidence for incomplete disclosure of information, as well as coercion and fraud.

Not all the fault is on the physician side of the relationship, but there is more power and therefore more temptation to abuse of power on the part of physicians.

There is evidence that patients act capriciously, irrationally, and out of their own self interests without regard for others. If one does not keep this aspect of the patient-physician relationship in mind then the proper checks and balances will not be structured into the clinical encounter.

CAUSE AND CORRECTION OF HUMAN FAULT

For Gustafson this human fault is descriptive of what Christian thought refers to as sin. "From a theological perspective the human condition is not only one of finitude, but also one of sin. There is a profound tendency in humans to secure narrow self-interests and the interests of particular communities at the cost of the well-being of others" ([9], p. 70).

The principle cause for the human fault is "our tendency to be turned inward toward ourselves as individuals, or toward our communal interests. It is . . . becoming more contracted in our being than we ought to be. It is a shrinking of self and community" ([10], p. 306). The origin of this "tendency to be turned inward" is a subject of theological debate, but it is best located in the nature of the human rather than any external cause.

Gustafson sees a potential for reversal or correction of the human fault when it is expressed as a disordered relationship which could be reordered by some ultimate ordering principle:

The human fault is not only moral, in the restricted sense of that term; it is a deeper misplacement of ourselves and our communities in relation to other persons and communities, and in relation to nature. Recall my general principle: we are to relate ourselves and all things in a manner appropriate to our, and their, relations to God. The human fault keeps us from proper understanding of our proper relations by contracting our trusts and loyalties, our loves and desires, our rational construing of the world, and our moral interests. The distortions that are forthcoming penetrate many areas of human experience and human action and they become pervasive. This is what the theological tradition has understood to be the corruption of man and of human life" ([10], p. 306).

The human fault can be characterized as a "contraction of soul" whose correction is an "enlargement of soul" through a reordering of relationships. "If the human fault can be indicated by the metaphor of the contraction of soul and on interests and its consequenes as improper relations of ourselves and all things to God, the correction can be indicated by an 'enlargement' of soul and interests, and by a more appropriate alignment of ourselves and all things in relation to each other and to the ultimate power and orderer of life. This is the correction that a theocentric view can evoke and sustain" ([10], p. 307).

Gustafson proposes that the correction process consists of three parts: "an alteration and enlargement of vision, which is in part a correction of the flaw of our rational activities; an alteration and enlargement of the 'order of the heart', which is in part a correction of the flaws of idolatry and of disordered loves and desires; and different standards for determining proper human being and action as a result of the other corrections, which is in part a correction of the flaw of 'disobedience'" ([10], p. 308).

Gustafson is aware that nontheocentric models have made some measure of success at reordering relationships. He persists in claiming that "the enlargement of vision that a theocentric perspective enables certainly can be achieved, at least in considerable measure, from nontheological perspectives. But the theocentric view does enlarge the context within which humanity is perceived and interpreted" ([10], p. 308).

THEOCENTRIC PERSPECTIVE

The corrective for human fault which Gustafson proposes is the theocentric foundation to his entire project of ethics. If human beings relate to all things in a manner appropriate to their relationship to God, then the well-being of the creation will be promoted. Whereas the contraction of the human spirit is the cause of the fault, enlargement of the human spirit is the correction of the fault. The correction is only partial but important.

The correction derives from a perspective which claims that there is an ordering power in the universe which provides a moral center for human conduct. "To make clear some aspects of a theological moral point of view the first two themes particularly must be kept in mind. God intends the well-being of the creation. God is both the ordering power that preserves and sustains the well-being of the creation and the power that creates new possibilities for well-being in events of nature and history" ([9], p. 27). The theological point of view which Gustafson is promoting is that God knows and wills that which is for the well-being of the creation.

Without the theocentric center to his ethics project, Gustafson would not be able to fully support his claim that human fault can be partially corrected so that human agents can promote the well-being of creation rather than their own well-being exclusively.

This central theme of Gustafson's system will not be the focus of this discussion. It clearly contrasts Engelhardt's intentional lack of reference to any transcendent dimension and his substitution of the humanum in the place of a God concept. While this contrast is to be kept in the background, my project is the comparison of the two systems in regard to the specific issue of human fault within the concept of persons.

LAKATOS' COMPARATIVE APPRAISAL METHOD

Which claims may count as scientific knowledge is constantly undergoing debate. What evidence is allowed as warrant for claims and how one set of evidence may be evaluated in relationship to a competing set of evidence is at the heart of the growth of knowledge. One method or another of comparative appraisal must be adopted in order to choose which set of claims is to be accepted and which is to be rejected by the community of appraisers.

This section will describe the method of comparative appraisal developed by the philosopher of science, Irme Lakatos. Once described, this method will be applied to a comparative appraisal of the competing theoretical claims of Engelhardt and Gustafson.

Although his description of the logic of discovery may seem arbitrary, Lakatos supports his methodology by historical examples of how scientific knowledge has grown over time. The method is the consequence of Lakatos' interpretation of the history of science specifically and human thought generally.

Lakatos' method allows for the comparison of two "research programmes" to appraise the degree to which one is more progressive or degenerative relative to the other. The method does not "prove" or "disprove" truth content of a programme. Instead, the method estimates the degree of acceptability of one programme relative to its rival.

Lakatos' method divides the scientific research programme into two parts: a hard core and a protective belt of auxiliary hypotheses. The hard core is a set of claims so central to the meaning of the research programme that it may not be questioned without negating the entire programme. These conceptual assertions have been determined by those who propose the research programme to be immune to questions of validation. The hard core is considered "irrefutable" by the methodological decision of its proponents. The choice of which elements of a theory are to be included in the hard core is not entirely arbitrary because the negative heuristic must represent very general and central dynamics of the theoretical position. Lakatos does not give any firm rules about the method for deciding which assertions should qualify for inclusion in the hard core. Whatever is placed there is "irrefutable by the methological decision of its proponents" ([14], p. 50).

The empirical claims contained in the hard core would usually be expected to be open to experimental examination. The unique move which Lakatos makes is to wrap a protective barrier around the core claims so that the core of the program is not the direct object of refutation. In his model the conceptual center of a theory is to be granted a certain temporary immunity from direct falsification so that it may remain undisturbed until it can be determined, by the dynamics of the positive heuristic, whether or not the research programme is progressive or degenerative.

The protective belt is a set of conjectures which can be tested by refutation. Whereas the claims of the hard core are immune to refutation, the auxiliary claims of the protective belt are the active site of refutation. It is the protective belt which is tested by experimental examination. "It is this protective belt of auxiliary hypotheses which has to bear the brunt of tests and get adjusted and re-adjusted, or even completely replaced, to defend the thus hardened core" ([14], p. 48). The procedures by which the auxiliary hypotheses are adjusted or replaced constitute the real action of theory competition.

Since the postive heuristic is the refutable portion of the research programme the rules of refutation must apply to its claims. Lakatos gives Karl Popper credit for having advanced the logic of scientific discovery by defining refutation in terms of falsifiability. Lakatos, however, translates falsifiability, or refutability, into "acceptability". This is an extremely important modification of Popper's method and it becomes the most active concept in Lakatos' comparative appraisal.

Lakatos furthers separates acceptability into three categories: "The Popperian scientist makes separate appraisals corresponding to the separate stages of discovery. I shall use these methodological appraisals (acceptability(1), acceptability(2), etc.) to construct an appraisal even of the trustworthiness of a theory, i.e., acceptability(3)" ([15], p. 170).

Acceptability(1) is a "prior appraisal of a theory" which follows immediately upon its proposal. It is by the dynamics of acceptability(1) that the "boldness" of a theory is appraised. Boldness for Lakatos has to do with the capacity of a theory to make predictions of novel fact. If the theory is bold in the sense that it entails some novel factual hypothesis, "scientists *accept(1)* it as a part of the body of science of the day" ([15], p. 171). This level of acceptability is prior to any testing. It is a type of acceptance in a scientific community which is conceptual and does not depend on empirical data.

Acceptability(2) relates to the empirical corroboration of the novel factual hypothesis proposed. In Lakatos' account: "bold theories, after having been severely tested, undergo a second 'posterior' appraisal. A theory is 'corroborated' if it has defeated some falsifying hypothesis, that is, if some consequence of the theory survives a severe test. It then becomes 'accepted(2)' in the body of science" ([15], p. 174).

These two methodological appraisals could be conducted without concern with the question of general truth or falsity of one theory in relation to another theory. All that is under consideration is whether or not enough empirical data can be found to corroborate some claim of one theory in excess over the claim of its rival theory.

Lakatos explores in acceptability(3) the question as to whether there can be any rational reconstruction of the logic of scientific discovery. He concludes that scientific growth is not merely a shift from one irrational (i.e., unproven) position to another based on the psychosocial forces at work within the community of scientists. The sociology of knowledge is based on the notion that social power determines which theoretical structure receives money for research, priority for publication, and students to be trained as disciples. In this way a paradigm theory establishes its dominance. This is what Lakatos identifies in Thomas Kuhn's model as "mob psychology" ([14], p. 178). On the other hand, he does not want to return to the claims of the justification-ists who demanded verification of theories according to some normative standards of final truth. Even though he rejects the idea of a standard of proof, Lakatos is interested in rehabilitating the rationality of science. He cannot reconstruct rationality without some standard of rationality to guide the reconstruction.

Lakatos believes that he sees a possibility of satisfying the need for an acceptability(3) in Popper's idea of "verisimilitude": "It seems obvious to me that the basis for a defintion of the intuitive idea of acceptability(3) should be Popper's 'verisimilitude': the difference between the truth-content and falsity-content of a theory. For surely a theory is the more acceptable(3) the nearer it is to the truth, that is, the greater its verisimilitude" ([15], p. 182).

How well Popper's notion of verisimilitude holds up in practical thought is an open question. The main point made by this concept is that probabilities are at work in all theoretical propositions and that it is not possible to discover with certainty the absolute truth-value of propositions. The key to knowing whether any theory lies closer to, or further from, the truth is its degree of verisimilitude, or "reliability": reliability not in the classical sense of an absolute, but in Popper's sense of an estimation.

Although originally designed with natural science in mind, Lakatos' method has been opened up to problems of social science. Lakatos initially claimed that his scientific research programme method could be applied to "any normative knowledge including ethics and aesethetics" ([14], p. 133). There has been a degree of application of Lakatos' method to the field of economics. The seminal work of Philip Hefner, Nancey Murphy, and Philip Clayton has demonstrated the utility of Lakatos' methodology in clarifying the comparative appraisal of theological theories [3], [12], [16].

Philip Hefner has been a leader in the movement to apply this comparative appraisal to the social sciences: "It is clear, therefore, that being right or wrong is not the only value of a research program, or even the chief value. Rather, the value lies in the program's fruitfulness in opening up constructive insights (new knowledge) to those who seriously subject its auxiliary hypotheses to discussion and possible falsification. This capacity to stimulate new insights, or the lack of such capacity, is the chief criterion for judging proposed theories" ([12], p. 25).

The criterion of fruitfulness can be interpreted along a spectrum of notions. One idea may confine fruitfulness within a strict limit of quantitative data accumulation. Another idea of fruitfulness may allow for any idea which can be pursued through mathematical constructs. Yet another idea of fruitfulness could include whatever promises to sell in the academic marketplace. Hefner has taken the position that fruitfulness must be defined as production of new insights.

I prefer to define fruitfulness more broadly to encompass "corroboration of old insights" as well as the "production of new insights". In the vigorous discussion among philosophers of science about what Lakatos meant, both of these notions of fruitfulness have strong and convincing proponents. It is this broader understanding of fruitfulness, or "acceptability", which will be applied in the comparative appraisal of Engelhardt and Gustafson.

In order to claim Gustafson's model as finally progressive over Engelhardt's model it is would be necessary for testing of this kind to be repeated through an extended series of auxiliary hypotheses. In actual historical scientific practice the series of comparative testing continues over long periods of time until one rival withdraws from the field of competition. The end of comparative appraisal is signaled when no theorist will champion the theory. Therefore, it is evident that this single comparative exercise performed here does not end the competition between these two rival theories. It is only one test episode in a lengthy series of tests. On other tests either of the two theories may be accepted as progressive over the other. Which theory will eventually be withdrawn is not predictable.

COMPARATIVE APPRAISAL

Lakatos' method requires that the auxiliary hypotheses in the protective belt surrounding the hard core be the aspect of rival theories which are tested by comparison. Therefore, the hard core statements about the theocentric and the humanistic grounding of the rival theories will not be directly involved in the comparison. Instead, the comparison will be located about the auxiliary hypotheses related to the competing concepts of person. The specific aspect of the concept of person will be the issue of human fault. The key to the argument is the following: if Gustafson's theory has a more acceptable construction in regard to this one issue of human fault, the entire theocentric model may be designated as being more progressive than Engelhardt's humanistic model. It is necessary to determine whether or not Engelhardt has a concept of human fault and how well developed that concept is in comparison to that of Gustafson.

There is some evidence that Engelhardt does recognize human fault. For example, he agrees that there are some individuals who will not qualify for being able to choose. Restrictive contracts, deception, threats of violence and other forms of coercion might limit a person's freedom of choice. ([5], p. 266).

Engelhardt comments about the instability of commitment to the will to morality: "One may indeed decide that, though humans are capable of understanding the difference between right and wrong and of pursuing the right for brief periods of time, they do not have the capacity for an enduring commitment to the good. If such is the case, it will not be an argument against the moral point of view, but only for the judgment that human history is intrinsically bound to evil" ([5], p. 384). If human history is intrinsically bound to evil it would seem appropriate for Engelhardt to take account of that in his theoretical construction.

Engelhardt goes so far as to say that "secular pluralist morality is thus founded on limits: the limits of reason and of authority" ([5], p. 386). This is a major limitation which makes it impossible to establish a single content-full canonical hierarchy of values. He is not so concerned with the faulted character of sinful human persons.

Engelhardt seems very close to opening up the idea of human finitude and fault in this passasge: "Secular humanism humanizes by introducing a sense of sophrosyne and a sense of human finitude. This has traditionally been an office of the humanities: to be an antidote to hubris" ([6], p. 139). In spite of this turn to the idea, Engelhardt does not pursue the consequences of finitude.

Engelhardt claims that "our whole being is fragmentary" ([7], p. 160). Because of "fragmentation" human beings may be in need of the remedy of grace: "The Judeo-Christian tradition has regarded human history as a conflict between the grace of God and the venality, cupidity, and inconstancy of human beings" ([5], p. 384). He affirms the effort of one of his theological teachers, Charles Hartshorne, to promote the heuristic value of a God's-eye view which acts as a counterforce to the fragmentary nature of the human being.

Despite all of this scattered evidence, Engelhardt does not have a developed idea of human vice or fault which can be identified as important to his ethics programme. Without a developed idea of human fault persistent problems remain unsolved in his ethics system.

One persistent problem is how to work out a peaceful settlement between conflicted positions when commitment to each position appears to preclude compromise. Compromise and integrity are not always compatible [1]. Persons are so strongly bound through interaction with their particular society that they are not at sufficient liberty to suspend the position of their moral community while they go about negotiating a settlement with a contrary moral position ([17], p. 563). It is likely that persons who hold a particular position will have to give up something of value within their own position in order to genuinely respect the contrary position of other persons. This is the nature of negotiation. Engelhardt is aware of this problem but does not offer a solution.

Another persistent problem is the actual nature of human beings as opposed to some ideal character. Persons, despite self-consciousness, rationality, and moral sense, are also flawed in character. A moral ideal in which one can trust persons to have a perfect, or even sufficient, will to morality does not exist in reality. Persons are self-deceptive, prone to lie to others, harbor self-interests, and are inconstant in their commitment to moral values. This strong trend in human nature is not fully appreciated in Engelhardt's programme of ethical conduct.

Engelhardt is not being realistic about the actual nature of clinical encounters as the interaction of persons who are not consistently self-conscious, rational, and moral. He has not satisfied the principle of miminal psychological realism suggested by Owen Flanagan: "Make sure when constructing a moral theory or projecting a moral ideal that the character, decision processing, and behavior prescribed are possible, or are perceived to be possible, for creatures like us" ([8], p. 32).

On the other hand Gustafson has a well developed idea of human fault as outlined above. Gustafson is realistic about human nature and conduct. He does not claim that human fault completely dominates human behavior. He demands responsibility for those acts which can be controlled. He also allows for the enlargement of human capacity to do good through the expansion resulting from the adoption of a theocentric perspective.

These features of Gustafson's theocentric model make it fruitful in understanding and guiding the human interactions entailed in the clinical encounter. Being more fruitful, the theocentric model should be more acceptable and, therefore, the more progressive in relation to Engelhardt's humanistic model.

CONCLUSION

Both of the competing ethics systems have strong features which recommend their acceptance by the community of science. Engelhardt has developed a convincing historical analysis in regard to the hope and failure of the enlightenment to create a rational consensus around a content-full moral vision. His compelling conclusion is that the fragmentation of moral communities into a pluralism has necessitated the creation of a neutral peaceable kingdom where moral differences can be negotiated by persons of good moral will. He has a consistent set of criteria for the concept of the persons who will create this neutral space and negotiate a satisfactory peace. This general pattern in human conduct can be applied to an analysis of the autonomy of persons in the clinical encounter.

With all of this construction Gustafson is in general agreement. The main differences are that Gustafson claims that the persistent presence of human fault significantly limits the entire moral project of negotiating a peaceful arrangement. Without some means of correcting the human fault, what Engelhardt promotes as necessary cannot be achieved. The human fault can be partially and tenuously balanced by adopting a theocentric perspective for all human action. "To relate to all things in a manner appropriate to their relation to God" is the corrective theocentric formula.

It has not proved fruitful in historical accounts of the battle between the theocentric and humanistic points of view to simply debate the central claims of the two models. An alternative comparative appraisal has been proposed by Lakatos through which rival theories can be tested for acceptability. This method can be applied to theories of the human sciences as well as the natural sciences.

By applying the comparative appraisal method in regard to the specific issue of human fault and the limits of the autonomy of person, Gustafson's theocentric model has been found to be more acceptable than Engelhardt's secular humanistic model. Therefore, the theocentric perspective can be recommended as being the more progressive model for ethics.

BIBLIOGRAPHY

1. Benjamin, M.: 1990, *Splitting the Difference: Compromise and Integrity in Ethics and Politics*, University of Kansas Press, Lawrence, KS.
2. Brody, H.: 1992, *The Healer's Power*, Yale University Press, New Haven.
3. Clayton, P.: 1989, *Explanation from Physics to Theology*, Yale University Press, New Haven, CT.
4. Engelhardt, H.T.: 1983, 'The Physician-Patient Relationship in a Secular, Pluralist Society', in E. E. Shelp (ed.), *The Clinical Encounter: The Moral Fabric of the Patient-Physician Relationship*, D. Reidel Publishing Company, Dordrecht, pp. 253–266.
5. Engelhardt, H.T.: 1986, *The Foundations of Bioethics*, Oxford University Press, New York.
6. Engelhardt, H.T.: 1991, *Bioethics and Secular Humanism: The Search for a Common Morality*, Trinity Press International, Philadelphia.
7. Engelhardt, H.T.: 1991, 'Natural Theology and Bioethics', in L.E. Hahn (ed.), *The Philosophy of Charles Hartshorne*, Open Court, La Salle, ILL, pp. 159–168.
8. Flanagan, O.: 1991, *Varieties of Moral Personality: Ethics and Psychological Realism*, Harvard University Press, Cambridge, MA.
9. Gustafson, J.M.: 1975, *The Contributions of Theology to Medical Ethics*, Marquette University Press, Milwaukee.
10. Gustafson, J.M.: 1981, vol. 1, 1984, vol. 2, *Ethics from a Theocentric Perspective*, University of Chicago Press, Chicago.
11. Gustafson, J.M.: 1991, 'Theological Anthropology and the Human Sciences', in S.G. Davaney (ed.), *Theology at the End of Modernity*, Trinity Press International, Philadelphia.
12. Hefner, P.: 1993, *The Human Factor: Evolution, Culture and Religion*, Fortress Press, Minneapolis, MN.
13. Katz, J.: 1984, *The Silent World of Doctor and Patient*, Free Press, New York.
14. Lakatos, I.: 1978, *The Methodology of Scientific Research Programmes*, in J. Worral and G. Currie (eds.), Cambridge University Press, New York.
15. Lakatos, I.: 1978, *Mathematics, Science and Epistemology*, in J. Worral and G. Currie (eds.), Cambridge University Press, New York.
16. Murphy, N.: 1990, *Theology in the Age of Scientific Reasoning*, Cornell University Press, Ithaca, NY.
17. Newman, L.E.: 1993, 'Talking Ethics with Strangers: A View from Jewish Tradition', *Journal of Medicine and Philosophy* 18, 549–567.

Notes on Contributors

Courtney S. Campbell, Ph.D., is Associate Professor, and Director, Program for Ethics, Science, and the Environment, Department of Philosophy, Oregon State University, Corvallis, Oregon.

Stephen E. Lammers, Ph.D., is the Helen H.P. Manson Professor of the English Bible, Department of Religion, Lafayette College, Easton, Pennsylvania.

Karen Lebacqz, Ph.D., is Professor of Social Ethics, McGill University, Montreal, on extended leave from Pacific School of Religion, Graduate Theological Union, Berkeley, California.

B. Andrew Lustig, Ph.D., is Academic Director at the Institute of Religion, and member of the Center for Ethics, Medicine, and Public Issues, Baylor College of Medicine, Houston, Texas.

Gerald P. McKenny, Ph.D., is Assistant Professor, Department of Religious Studies, Rice University, Houston, Texas.

Timothy E. Madison, Ph.D., is Director of Pastoral Services, Valley Baptist Medical Center, Harlingen, Texas, where he is co-chair of the Bioethics Committee.

Michael M. Mendiola, Ph.D., is Assistant Professor of Christian Ethics, Pacific School of Religion, Graduate Theological Union, Berkeley, California.

Robert Lyman Potter, M.D., Ph.D., is Clinical Ethics Scholar, Midwest Bioethics Center, and Associate Clinical Professor of Medicine, Kansas University School of Medicine, Kansas City, Missouri.

Thomas A. Shannon, Ph.D., is Professor of Religion and Social Ethics, Department of Humanities and Arts, Worcester Polytechnic Institute, Worcester, Massachusetts.

Earl E. Shelp, Ph.D., is Executive Director, Foundation for Interfaith Research and Ministry, Houston, Texas.

Paul D. Simmons, Ph.D., teaches ethics at the University of Louisville, and was Professor of Christian Ethics, The Southern Baptist Theological Seminary, Louisville, Kentucky.

David C. Thomasma, Ph.D., is the Fr. Michael I. English, S.J., Professor of Medical Ethics, and Director, Medical Humanities Program, Loyola University Chicago Medical Center, Chicago, Illinois.

Robert M. Veatch, Ph.D., is Director, Kennedy Institute of Ethics, Georgetown University, Washington, D.C.

Gerald R. Winslow, Ph.D., is Dean, Faculty of Religion, Loma Linda University, Loma Linda, California.

Index

Theology and Medicine

Managing Editor

Earl E. Shelp, *The Foundation for Interfaith Research & Ministry, Houston, Texas*

1. R.M. Green (ed.): *Religion and Sexual Health.* Ethical, Theological and Clinical Perspectives. 1992   ISBN 0-7923-1752-1
2. P.F. Camenisch (ed.): *Religious Methods and Resources in Bioethics.* 1994
                                                                                  ISBN 0-7923-2102-2
3. G.M. McKenney and J.R. Sande (eds.): *Theological Analyses of the Clinical Encounter.* 1994   ISBN 0-7923-2362-9
4. C.S. Campbell and B.A. Lustig (eds.): *Duties to Others.* 1994   ISBN 0-7923-2638-5
5. E.R. DuBose: *The Illusion of Trust.* Toward a Medical Theological Ethics in the Postmodern Age. 1995   ISBN 0-7923-3144-3
6. L.S. Cahill and M.A. Farley (eds.): *Embodiment, Morality, and Medicine.* 1995
                                                                                  ISBN 0-7923-3342-X
7. J.B. Tubbs, Jr.: *Christian Theology and Medical Ethics.* Four Contemporary Approaches. 1996   ISBN 0-7923-3657-7
8. E.E. Shelp (ed.): *Secular Bioethics in Theological Perspective.* 1996
                                                                                  ISBN 0-7923-3735-2

KLUWER ACADEMIC PUBLISHERS – DORDRECHT / BOSTON / LONDON

DATE DUE

FEB 0 2 1998			
DEC 1 9 2001			
MAY 0 2 2005			
APR 2 4 2009			